Sand, Science, and the Civil War

Sand, Science, and the Civil War

Sedimentary Geology and Combat

SCOTT HIPPENSTEEL

The University of Georgia Press
Athens

© 2023 by the University of Georgia Press
Athens, Georgia 30602
www.ugapress.org
All rights reserved
Set in 9.75/13.5 Baskerville 10 Pro Regular
by Kaelin Chappell Broaddus

Most University of Georgia Press titles are
available from popular e-book vendors.

Printed digitally

Library of Congress Cataloging-in-Publication Data

Names: Hippensteel, Scott, 1969– author.
Title: Sand, science, and the Civil War : sedimentary geology and combat /
Scott P. Hippensteel.
Other titles: Sedimentary geology and combat
Description: Athens : The University of Georgia Press, 2023. | Series: Uncivil wars | Includes
    bibliographical references and index.
Identifiers: LCCN 2022036129 (print) | LCCN 2022036130 (ebook) | ISBN 9780820363530
    (paperback) | ISBN 9780820363523 (hardback) | ISBN 9780820363578 (pdf) |
    ISBN 9780820363547 (epub)
Subjects: LCSH: United States—History—Civil War, 1861–1865—Campaigns. | Military
    geology. | Landscapes—United States—History—19th century. | Sediments (Geology)—
    United States—History—19th century. | United States—History—Civil War, 1861–1865—
    Environmental aspects. | United States—Strategic aspects. | Southern States—Strategic
    aspects.
Classification: LCC E470 .H67 2023 (print) | LCC E470 (ebook) | DDC 557.3—dc23/
    eng/20221012
LC record available at https://lccn.loc.gov/2022036129
LC ebook record available at https://lccn.loc.gov/2022036130

CONTENTS

ACKNOWLEDGMENTS

The general concept for this work was framed more than twenty years ago as I spent countless hours slogging through the marshes of coastal Delaware. The wetlands contained more than the sedimentary strata I was sampling for my graduate research—they were also surrounded by World War II-era gun batteries and shoreline reconnaissance towers. It was here, studying at the University of Delaware, that I learned a great deal about sedimentary geology and coastal processes from my two most influential professors of geology: James Pizzuto and John Wehmiller. It was also at Delaware where I met M. Scott Harris, a fellow graduate student, who I am convinced knows more about the geology of the Coastal Plain than any other person on the planet. I learned a great deal from him as well while working on the sediment samples from the *H. L. Hunley* submarine.

At the other end of the time line for this project, two other scholars at a different university also helped me complete this book. Mick Gusinde-Duffy, the executive editor at the University of Georgia Press, provided encouragement and much-needed guidance for how best to work my way through the 537 peer reviews that accompanied the publication process (perhaps it wasn't quite this many, and I concede that the anonymous reviewers provided critically important feedback that greatly improved the manuscript). Stephen Berry, the coeditor of the *UnCivil Wars* series, assisted me greatly in organizing and crafting the book to find a proper path between the natural and the military sciences.

Finally, as always, I would like to thank my wife and daughter for their support and inspiration. The creation of this book would not have been possible without their patience and encouragement during the many years I spent writing, reorganizing, and revising this manuscript.

A NOTE ABOUT
UNITS OF MEASUREMENT

In this book the references to distances and weights that are in the historical realm are listed in imperial units; those that deal with scientific measurements are given in metric units (e.g., Pickett's Charge covered half a mile of open ground, while medium-fine sand is 0.25 mm in diameter).

Sand, Science, and the Civil War

PART I

Sedimentary Geology and Warfare

With mediocre troops one must shift much soil.
> —Napoleon Bonaparte

A shell burst in the pit in which I was standing with the men, covering us with a column of sand, without injury to any of us.
> —John Porter Fort, *Memoirs of the Battle of Bentonville*, 1903

... the penetration of rifle projectiles, fired into a sand parapet ... is but trifling.
> —Maj. Gen. Quincy Gillmore, report from
> Morris Island, South Carolina

CHAPTER 1

Sediments and Conflict

The Nature of Sediments

Sediments, in the form of clay, silt, sand, pebbles, and cobbles, are as limitless in supply as they are versatile in function. Depending on their size, sediments can behave in a strikingly discordant manner. For example, some sediments are highly porous, allowing water to pass through them quickly; others are completely impermeable. Adding water to some tiny sediments makes them strongly cohesive, yet adding just a little more liquid to the same wet particles renders them exceedingly slippery. Coarse fragments of rock can be spread to make a roadway more traversable, while much finer grains might render a road impassable.

This book is about how these sediments and the dichotomous nature of the detrital particles shaped the combat, strategy, and logistics of the American Civil War. The intersection between sedimentary geology and military history has received meager attention in the literature, yet the influence of sediments on the conduct of the war was profound.[1]

The effect of the sediment on the combatants was determined by the grain size and composition of the collective particles. On the defensive, any type of sediments could be piled to provide soldiers with concealment and cover; during the same fight, sediments might be exploited on the offense, using fragments of rock, or clasts, as weapons. At Second Manassas, for example, when members of the Stonewall Brigade ran low of ammunition, they—fittingly—threw cobbles with concussive results. Boulders were rolled down the side of Kennesaw Mountain in combat. At Petersburg, handfuls of sand were tossed into the faces of the oncoming enemy as a means of temporarily blinding a close-quarters foe.

Sediments can make weapons more deadly or undermine their damaging effects. If an exploding artillery shell hits very coarse sediments, a new rocky shrapnel will result, but if the same shell strikes fine-grained sediment, the exploding heavy ordnance can suddenly be rendered nearly harmless. Fine, penetrating sediment can also make a rifle musket as useful in combat as a nine-pound club.

Some sediments were discovered to help heal wounds, while other identical sediments carried anthrax. Sediments were occasionally used for camouflage—painting the sides of a ship with dark mud prior to night combat, for example—but changes to the color of a sediment might also be an indicator of danger—like the lighter-colored disturbed sand above a recently buried torpedo (land mine).

The dual nature of the sediment even influenced the soldiers on the march. When large volumes of water were added to the sediment, soldiers' boots, cannon, mules, and wagons would sink, along with their morale, while the drying of the same sediment resulted in a second, even more loathed, airborne irritant: dust.[2]

With all of these influences, the key determinant was the sediment's size.[3] Large sediments were stacked by the *thousands* at Gettysburg to build fortifications or breastworks. Smaller sediments were excavated in the *trillions* to produce the trenches of Vicksburg and Petersburg and piled in the *quadrillions* along the coastlines to turn beaches into fortresses.[4] The sediment that was shoveled from the beaches and dunes and piled into massive sandcastles along the shore was, of course, sand. It was this granular material, with a grain size ranging between 0.06 and 2.0 millimeters in diameter, that had the greatest impact on the fighting of the Civil War.

Sediment and Terrain

Terrain is a function of geology and climate. Optimal terrain could be exploited by a skilled commanding officer both on offensive maneuvers and during defensive positioning of troops and artillery. This book explores how the landscape came to be shaped the way that it is (geomorphology) and how, on a range of scales, sedimentary geology influenced the fighting of the war.

Studies of Civil War tactics and terrain are numerous, and those connecting terrain, geology, and fighting exist for several major battles.[5] The majority of these studies have concentrated on the Battle of Gettysburg, and for good reasons. The Adams County, Pennsylvania, landscape surrounding Gettysburg is distinctive when compared to other battle sites from the Eastern and Western Theaters of conflict. There are few places where the geology of a battlefield and its influence on combat can more easily be imagined than the square mile of land below Little Round Top (see figures 1.1 and 1.2).[6]

A second reason that Gettysburg's geology has received so much attention from researchers is the size and importance of the battle. No other battle involved as many troops or had more casualties, and had Lee's army been able to break the center of the Union line during Pickett's Charge, the path to Harrisburg, Baltimore, Philadelphia, or Washington might have been open.

The most famous rocks in the Gettysburg Basin are of the igneous variety; nevertheless, the hard rocks that underlie the Round Tops, Devil's Den, and Cemetery and Seminary Ridges would not have been as notable without the presence of the surrounding sedimentary rocks. The harder rock's expression on the landscape simply would not have been as significant without the more rapid and consistent weathering of the adjacent softer sedimentary rocks.

This phenomenon of variable rates of erosion is called *differential weathering*. The weathering characteristics and stratigraphy of the local bedrock determines the relief (slope) of a battlefield.[7] At Antietam and Chickamauga, for example, different varieties of carbonate rock weathered at different rates, producing a landscape with hard(ish) rock ridges and rolling hills. Softer, more easily erodible limestones and shales created swales and valleys.[8]

All of these differences in how a rock is destroyed create interesting military engineering challenges for soldiers of all ranks and also determine what type of sediment might be present on the battleground. Sediments dictated both where fortifications could be constructed and what materials were available to convert the landscape into an elaborate series of ditches, trenches, and parapets. Different campaigns were waged on landscapes underlain by a large variety of sediments and sedimentary rocks. The influence of sedimentary geology is explored in these pages with respect to visibility (sight lines and concealment), tactics (entrenching and tunneling), and strategy.

FIG. 1.1. A version of this photograph, taken by T. H. O'Sullivan in the days immediately after the fighting in Devil's Den, was titled "Rocks could not save him at the Battle of Gettysburg." In 1975 the photograph was demonstrated to be staged by a local historian, William Frassanito. Library of Congress.

FIG. 1.2. A second photograph from near Devil's Den. This photograph, taken by Alexander Gardner, shows Confederate dead from the "Slaughter Pen," an area of outcrops at the base of Little Round Top. This photograph was taken approximately 25 yards east of the dead soldier pictured in FIG. 1.1. Library of Congress.

Preliminary Military Lessons about Sand

On July 18, 1863, Quincy Gillmore observed as the Union army and navy sent cross fire of nearly ten thousand heavy artillery shells into a sand fortification named Battery Wagner on Morris Island, South Carolina. Confident of the nearly total reduction of the fort and its eighteen-hundred-man garrison, the commanding officer ordered a five-thousand-man infantry assault to capture the disabled battery. To the great dismay of the Federal infantry and their leadership, however, what appeared to be a battered fort with a wounded garrison was instead a fully functional fortification with a nearly complete defensive capability. The ten-hour bombardment, at times raining shells at two-second intervals, had killed only eight men and wounded only twenty.[9] Gillmore's grand assault on Battery Wagner turned disastrous for the Union, who suffered 1,515 casualties during the failed assault (more than 30 percent of the attacking force).[10] The outnumbered Confederate defenders, sheltered in the sand, lost only 222 men.

The surprising resistance of Battery Wagner to reduction from bombardment, and the failure of the Union assault, can be directly related to the construction materials chosen for the fortification: medium quartz sand. It was on islands like Morris where sediments turned out to be the greatest force multiplier during the Civil War. Earthen and sand fortifications, including Battery Wagner, Fort Fisher, and Fort McAllister, proved largely indestructible, even under the fire of the newly evolving rifled artillery—the same modern guns that were concurrently turning brick fortifications into rubble.[11] This book details why sand was so beneficial to defending troops along the coastline. Additionally, sediments of other grain sizes (e.g., clay and silt), and other types of sedimentary rocks, are considered with respect to their use in fortifications and influence on transportation and logistics during the war.

The evolution of weapons and tactics between 1861 and 1865 was increasingly influenced by sedimentary geology. Entrenching, for example, was relatively uncommon during early battles in the war because the commanding officers thought digging in would diminish the offensive aggressiveness of their soldiers.[12] By 1862, both armies were taking full advantage of preexisting defensive cover (i.e., the famous Sunken Road at Antietam or the Stone Wall below Marye's Heights at Fredericksburg). From 1863 onward, both sides were entrenching as often and as quickly as the geology would allow. On battlegrounds underlain by unconsolidated sediments (or weakly cemented sediments) like Vicksburg and Petersburg, complicated earthworks and forts were constructed for the protection of strategically important sites like cities, bridges, and railroads. While the sediments allowed soldiers and sappers to rapidly construct defensive works that were highly resistant to attack, the same clastic particles made the fortification vulnerable to attack from below, making mining an increasingly common offensive tactic.[13]

This text is organized to effectively synthesize geology, military engineering, and history, with a running commentary on how the three fields were allied. The first section of the book (part I), including the first four chapters, discusses the intersection of geology and history. The second chapter discusses the work of two eminently important scientific influencers from the nineteenth century: the geologist Charles Lyell and the military engineer and theorist Dennis Mahan. Their work evolved at almost exactly the same time prior to the outbreak of the Civil War, and the link between their fields of study has never been properly explored. Chapter 2 also makes the case for why sedimentary rocks, out of all categories of earth materials, were the most important with respect to the combat and tactics from the war. One key component of this discussion involves the physical characteristics of a sediment or rock that made it preferable for use in Mahan's fortifications and fieldworks. Sandbags were employed in the tens of thousands during numerous Civil War campaigns, and there is a reason uncountable sandbags are still employed around security strongpoints and military bases today.

Chapter 3 considers the evolution of artillery and how the changing ordnance affected the construction (and destruction) of sedimentary field fortifications (chapter 4). The introduction of new heavy rifled field and siege artillery seemed to portend the end of coastal fortifications until the first shell hit a sand parapet with unexpectedly suppressed results. As historian James McPherson eloquently simplified: "[the sand parapets] absorbed shot and shell as a pillow absorbs punches."[14]

The remaining parts of the book continue the discussion of how history and geology interact, starting with harder rocks from the Piedmont and Valley and Ridge physiographic province (part II) and winding its way first to the softer rocks of the Coastal Plain (part III), before turning west to the Mississippi River Valley (part IV). Interspersed throughout these geographically demarcated sections of the book are additional chapters discussing the influence of sedimentary geology on transportation and logistics, as well as how sediments affected morale. After flowing through the Mississippi River Valley and discussing the Vicksburg Campaign, the text returns to where sediment had the most profound defensive influence during the war: the sandy coastline (part V). Here the massive sandcastles that defied the Federal army and navy are discussed in detail. The final section of the book, part VI, delineates the lessons that were carried out of the Civil War by military theorists and

FIG. 1.3. Karrens, or "cutters" were used to great effect as natural breastworks by Union infantry during the Battle of Stones River on December 31, 1862.

practitioners for later, greater wars. These final chapters also discuss how sediments can be used as a tool for historians and what the state of preservation is for many of our sedimentary fortifications from the 1860s.

More often than not during the Civil War, the high ground that proved so important for defensive stands was created by the weathering of sedimentary rock. Sedimentary rock also was a defensive force multiplier in a vast array of other, less well understood ways. The sand forts encountered by Quincy Gillmore were a nasty surprise, of course, with the small sediment benefiting the Confederate defenders on a large scale. In contrast, the defensive stand of the Union army at Stones River was greatly aided by the natural limestone trenches, or karrens, that cropped out along the center of their line (see figure 1.3).

July 1863

Federal and Confederate armies were fighting everywhere in July of the third year of the war. Along the Mississippi River, Ulysses Grant and John Pemberton were battling in the never-ending trenches that were easily cut into the soft, semi-cemented silt around Vicksburg. In the

Eastern Theater, George Meade and Robert Lee were struggling across hard-rock sandstones and siltstones at Gettysburg. In yet another theater, Quincy Gillmore was suffering great losses against the sand fortress of Battery Wagner on Morris Island. These three great battles, all within days of each other, witnessed very dissimilar tactics and had vastly different outcomes. Sediments and sedimentary rocks shaped these tactics and influenced the results of these campaigns, *but in completely different ways for each engagement.*

The dichotomous nature of the sediment, based largely on grain size, had profound effects on the fighting at all scales. This book examines this intersection of geology and history, using the terrific variety of Civil War landscapes as a lens through which to explore the role of sediments and sedimentary rocks on the combat and tactics of the war.

CHAPTER 2

The Coevolution of Military and Geological Sciences

Mahan and Lyell

The most influential professor of military science prior to, and during, the Civil War was Dennis Hart Mahan.[1] While at West Point, he wrote extensively about the design of permanent, temporary, and field fortifications, as well as the proper application of tactics and strategy. Much of his work also focused on the potential offensive and defensive exploitation of terrain and topography. Mahan's most influential work, *Treatise on Field Fortifications*, was published in 1836 and became the standard textbook for American military planners and engineers during the mid-nineteenth century.

Within only a few years of the publication of Mahan's *Treatise*, another great influencer, the Scotsman Charles Lyell, published the first two textbooks on geosciences: *Principles of Geology* (1833) and *Elements of Geology* (1838).[2] These books outlined, for the first time, how and why sediments were created and deposited and how the landscape came to be shaped to resemble the terrain and topography described in Mahan's work.[3] Lyell had a particular research interest in one type of silty sediment called loess, a material that would later become an important natural feature throughout the Vicksburg Campaign. Lyell would also travel to America, seeking geological examples from the antebellum South for his textbooks.[4] During one voyage, he traversed the soon-to-secede states from the Atlantic Ocean to the Mississippi River basin, crossing much of the ground that would later be contested during Sherman's campaign against Atlanta.[5]

By the start of the Civil War, the ideas brought forth by the engineer Mahan and the geologist Lyell were in their infancy, so neither discipline of natural or military history had time to consider the interaction between the sciences.[6] This intersection between fighting and sedimentology would have to wait until the third year of the war, when another military engineer, Union brigadier general Quincy Gillmore, would begin to conduct scientific tests on a geological material that was causing his men no end of frustration: pure, crystalline, quartz sand.

The Variety of Rock Types Underlying Civil War Battlefields

Geology was the single largest determining factor with regard to the lay of the land for Civil War battlegrounds. The geomorphology of the battlefields was influenced by geology and climatology, but the differences in climate between battlefields across the central and eastern United States are relatively inconsequential compared with the vast array of rock types found on the continent.[7] In other words, the differences in landscape features between Gettysburg, Pennsylvania, and Vicksburg, Mississippi, are much more strongly related to geology than temperature and precipitation.

Sedimentary rock and sediments are found under the largest proportion of Civil War battlefields (see figure 2.1). These rocks can be subdivided into three categories: unconsolidated sediments, lithified clastic sedimentary rocks, and carbonates.[8] Unconsolidated sediments, like sand and gravel, almost always result in the flattest terrain where entrenchments tended to be more easily dug. These battlefields are found almost exclusively on the Coastal Plain, including those around Richmond and Petersburg. One of the last large-scale fights to occur during the Civil War, the Battle of Bentonville in eastern North Carolina, also took place on the flattest, sandiest landscape (see figure 2.2). When buried, compressed, and cemented together, these clastic particles become lithified sedimentary rocks. Another type of sedimentary rock includes chemical carbonate rocks, such as limestone or dolostone.[9] These rocks often weather to produce interesting landscapes because of their solubility, especially in regions with slightly acidic precipitation.

Igneous rocks, which formed (crystallized) from molten magma, can be divided into two categories: ancient, immense igneous bodies and slightly younger igneous intrusions. Many of the widespread igneous

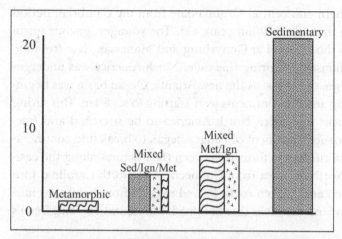

FIG. 2.1. Sedimentary rocks underly the majority of the major Civil War battlegrounds. This plot compares the underlying geology at the thirty largest battles from the Civil War, as defined by most soldiers present on the battlefield.

FIG. 2.2. One of the flattest, if not the flattest, battlefields from the Civil War: Bentonville. The battlefield is located on the western edge of the Coastal Plain. Had the battle taken place only twenty miles to the west, there would certainly have been some degree of slope or relief. This is the view from the Union Artillery position on the Morris Farm.

rocks in northern and central Virginia date from the Cambrian period and are more than 500 million years old. The younger igneous intrusions, such as those found at Gettysburg and Manassas, date from the Triassic and Jurassic.[10] During this time, North America was undergoing a great degree of stress, as the new Atlantic Ocean basin was beginning to develop and the continents were starting to separate. This rifting caused the rocks in eastern North America to be stretched and fractured.[11] As the supercontinent of Pangea began to break into continent-size fragments, rift basins formed between the fractures along the eastern coast of North America from Connecticut to North Carolina. Into these weakened and broken rocks flowed magma from below, eventually cooling to form the rocks that became Little Round Top and Devil's Den.[12]

Metamorphic rocks have been altered by heat, pressure, or chemical reactions associated with superheated fluids flowing through deeply buried rocks. As with the other groups of rocks, metamorphic rocks on Civil War battlefields can be subdivided into groups: on a large scale, entire regions of rocks can be altered through burial and increased pressure or during continental collisions. This is the case with the rocks around Chancellorsville, Spotsylvania Court House, and the Wilderness in Virginia. On a smaller scale, intruding 2,000°F magma can bake rocks with which it comes into contact. The railroad cut that played an important role during the first day of fighting at Gettysburg contains rocks that exhibit this "contact" metamorphism. The igneous intrusion that produced Seminary Ridge baked the older sedimentary rock it was cutting through.[13] The result is a metamorphosed baked zone in the sandstones and shales adjacent to the igneous rock that was later exposed when the construction of the railroad bed excavated into the ridge (see figure 2.3).

Some battlefields are underlain by a mixture of rocks, a combination that can produce interesting, and tactically important, terrain features. Gettysburg is, of course, the classic example of mixed geology producing good defensive positions. Igneous intrusions underlie Seminary and Cemetery Ridges, Culp's Hill, and the Round Tops, while the flatter open ground is underlain by more easily erodible (and farmable) sedimentary rock.

Richmond, Petersburg, and Fredericksburg, Virginia, have similar and interesting geology. All three are located on or very near the Fall Zone—the contact between the sedimentary rocks and sediments from the Coastal Plain and the harder igneous and metamorphic rocks of the

FIG. 2.3. Contact metamorphism at Gettysburg. Hornfels is in the foreground (left), with diabase in the background (right). As the diabase intrusion moved towards the surface it baked the surrounding sedimentary strata, resulting in the hornfels—a zone of metamorphosed rock surrounding the dike.

Piedmont. The term "Fall Zone" originated from the waterfalls and rapids associated with the rivers flowing across the contact area.[14] When traveling upriver by boat, it is necessary to disembark just below the rapids; the result was an ideal location for commerce and transportation hubs, and thus towns and cities developed at the base of these falls.[15]

In Richmond, for example, the rocks and sediments east of the rapids on the James River are very different from those to the west. All battles from the Peninsula Campaign took place east of the city on the softer, partially lithified sandstones and sediments of the Coastal Plain. Petersburg is located just to the south of Richmond and has a similar geology. The mineshaft that was dug to explode the famous Crater at Petersburg would never have been possible had the shaft been excavated west of the Fall Zone instead of in the sands and clays of the Coastal Plain.

Clastic and Chemical Sedimentary Rocks

Sediments are the weathered products of preexisting rocks. These clastic particles have a large range in particle (grain) size and are classi-

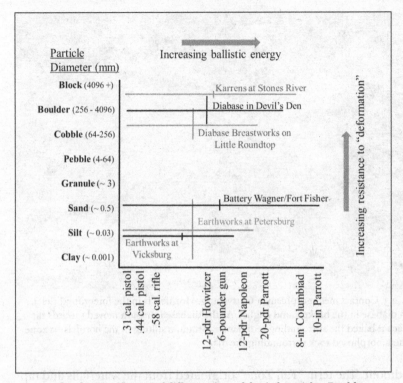

FIG. 2.4. Combat significance of different sizes of detrital particles. Boulders could defeat artillery rounds, small boulders and large cobbles could deflect small-arms fire, and sand could be piled for strong, artillery-resistant parapets. Vertical bars represent variations in grain size for a particular geological feature.

fied according to their diameter. Smaller grain sizes such as clay, silt, or sand are of most interest when building earthworks or fortifications while larger grain sizes such as cobbles or boulders were more useful for breastworks (see figure 2.4).

Sediments remain unconsolidated until they go through the process of lithification. This most often occurs when sediments are buried and compressed and mineral-rich groundwater cements the grains together. Collectively these grains of sand and silt become sandstone and siltstones, and the detrital (mechanically broken-down) particles now make up part of a sedimentary rock with a clastic texture.

The composition (mineralogy) and texture (grain size, roundness, sorting) of a sediment often provide clues about the origin of the sediment and the distance the particles have been transported. For example, a beach sand is typically composed of fine/medium-size quartz sand grains (around 0.5 mm in diameter) that are well rounded and well sorted.[16] The consistent grain size (few pebbles and little clay) and composition (few dark mineral grains: quartz is more chemically durable than darker and denser minerals such as pyroxene or olivine) is attributable to the distance the particles have traveled during the process of erosion.[17] These sediments are likely very far from their original source, and as they traveled by wind and water they have been tumbled, sorted, and smoothed. A river, for example, will bounce the particles together, smoothing rough edges, and the water current will leave larger particles behind and carry clay into the ocean or onto floodplains. The resulting sediment at the beach is composed of a single sandy grain size with well-rounded particles, a type termed "pure" or "clean" by geologists.[18] Also, with little silt or clay mixed in the beach sand, the sedimentary texture is well sorted and described as "mature." Note that maturity is determined by distance of transport, not the age of the sediment.

Chemical sedimentary rocks have a second type of sedimentological texture. These rocks are created from material that was originally dissolved in water before it precipitated out of solution through organic and inorganic processes. Creatures with shells or tests composed of calcium carbonate ($CaCO_3$), such as clams, sea urchins, and microfossils like foraminifers, contribute greatly to the construction of chemical sedimentary rocks like limestone.[19] The most common group of chemical, or dense, sedimentary rocks are the carbonates. These include dolostones ($CaMg(CO_3)_2$) and limestones ($CaCO_3$). Limestone is slightly softer than dolostone but more common. It represents approximately one-tenth of the volume of all sedimentary rocks and underlies about one-third of all Civil War battlegrounds.[20] Both limestone and dolostone can form in a variety of depositional environments, ranging from coral reefs to deep-marine basins. The one characteristic of nearly all of these depositional environments is that they are inhabited by creatures that use calcite to construct their skeleton, test, or shell (or in the case of deepwater environments, they are found below environs with swimming or floating calcareous critters).

Carbonates were important during the Civil War not because of how they were formed but because of how they are destroyed. Because these rocks were composed of material that was originally soluble, such as calcium carbonate, they are not particularly resistant to being dissolved again. In the field most geologists carry a small bottle of diluted hydrochloric acid for the purposes of testing the composition of rocks: car-

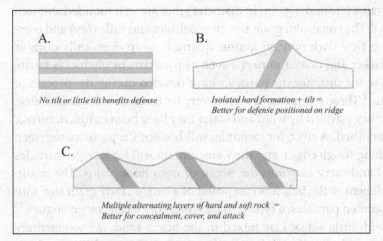

FIG. 2.5. Cross section illustrating the relationship between carbonate rock stratigraphy, differential weathering, and the resulting surface expression that could be exploited by commanders employing either offensive or defensive tactics. Darker grey represents more durable carbonates. Note that "B" and "C" offer the benefit of a reverse-slope firing position.

bonate rocks will effervesce under a drop of acid, but most clastic rocks will not. Limestone fizzes vigorously, dolostone less so, and sandstone shows no reaction unless the cement holding the sand together is composed of calcite.

As a result of this chemical instability and lack of durability, different limestones and dolostones will weather at different rates. Depending on the stratigraphy (layering and tilt of the rock units) and variety of rock types, the resulting landscape can have a variety of terrain features that were important for fighting soldiers. For the most part, limestone weathers more quickly than dolostone because it isn't quite as hard.[21] If either rock is enriched or interbedded with chert, another chemical rock composed of durable microcrystalline quartz, it will be much tougher and resistant to erosion. The resulting carbonate landscape, then, will be determined by the type of rock, the tilt of the rock layers, and whether any of the rock units contain harder minerals. In general, dolostones or chert-enriched carbonates produce ridges, and relatively thinly interbedded and tilted limestones and shales, alternating with dolostones, will produce rolling landscapes (see figure 2.5).

The Antietam battlefield (chapter 6) demonstrates the impact of differential weathering well. The morning phase of the battle was fought on the northern part of the battlefield, where the geology was most similar to the cross-section in figure 2.5A. Union casualties were particularly high when conducting the morning assaults across this flat landscape.[22] During the afternoon phase of the battle, the attacking Federal soldiers had more success attacking the Confederate center, underlain by strata similar to figure 2.5C. Despite having the benefits of the famous Sunken Road (later "Bloody Lane"), the Confederate defenders could not sustain their defensive stand in the rolling terrain.

This book describes the influence of sedimentary rocks and sediments on the tactics employed by both armies during the Civil War and the effects of these materials on the weapons, fortifications, and landscapes from the conflict. Whether of the clastic or carbonate variety, sedimentary geology and sedimentary rocks were important on far more battlefields than either igneous or metamorphic rocks, and their influence on terrain and combat has been underappreciated by historians.

CHAPTER 3

Killing at Range

Artillery and Geomorphology

Evolving Ordnance

The U.S. Ordnance Department, which is responsible for providing weapons to the army, has been notoriously conservative when it comes to utilizing evolving technology.[1] In the nineteenth century much of the leadership in this branch of the military was represented by elderly military officers who clung to the "proven" weapons from earlier wars in which they had participated. These officers also tended not to trust the discipline of common soldiers regarding ammunition management in combat. Examples of this conservativism with respect to new weapon technology, and its detriment to the fighting army, are not hard to find in the country's relatively young military history: Marines entering combat in World War II with bolt-action rifles that were standard technology during the previous worldwide conflict, instead of the semiautomatic rifles available at the time, or George Armstrong Custer's cavalry at the Little Big Horn armed with single-shot Springfield Trapdoor carbines when at least some of their foe were armed with repeating rifles. Even the scarcity of breach-loading rifles or repeating Spencer and Henry rifles during the Civil War were testament to the reluctance of the Ordnance Department to experiment with something new and perhaps battle-changing.

This same reluctance to employ newly developed small arms and equipment applied equally to the artillery used during the Civil War. Smoothbore artillery had been evolving at an extremely slow pace since the American Revolution. Smoothbore field guns and howitzers cre-ated a manageable amount of stress within the barrel because the seal between shot and barrel was not particularly tight. While this unfortu-nately limited the range and accuracy of the weapon, it did allow for a lighter barrel and a (slightly) more rapid reloading period. Similarities between smoothbore artillery and smoothbore muskets are not hard to see.

During the decade prior to the Civil War, the British were experi-menting with and implementing new types of rifled artillery. Rifling a barrel requires the cutting of grooves into the bore, and these grooves rotate the projectile when fired, increasing both the stability of the shell and the range of accurate fire. The tight seal between shell and the rifled bore requires more gunpowder to propel the projectile and thus more pressure from within the base of the barrel. More pressure equates to more range but also more danger from a bursting barrel, or the neces-sity of a heavier, reinforced barrel.

At the outbreak of the war, most artillery in the U.S. arsenal was de-signed for seacoast defense or heavy siege operations. Military plan-ners saw the coastline as the first line of defense, and they foresaw an attack from foreign powers (and navies) to be a more probable threat than an attack from within that would require significant quantities of field artillery.[2] With the advent of the rebellion, this situation regarding the scarcity of field artillery clearly favored the industrialized North. Northern factories and foundries could rapidly switch to the produc-tion of modern cannon. The South was left with what could be seized from local southern fortifications and armories, and only a single can-non foundry, the Tredegar Iron Works in Richmond, could produce field artillery at the time.

This chapter explores the variety of weapons used by both armies (and navies) during the war and how the effectiveness of these weap-ons evolved and was influenced, and in some cases greatly diminished, by sedimentary geology. Sand and sediments affected exploding ord-nance in very different ways, depending on the type of gun and the na-ture of the target. How much, for example, was the Union's artillery su-periority along the coastline minimized by the Confederacy's rapidly constructed earthen fortifications? Was the South's inability to con-struct elaborate masonry fortifications during wartime actually a ben-efit because of changes in artillery performance over time? Had they been given more time and resources, might the Confederacy have constructed more vulnerable brick fortifications at the (future) sites of Battery Wagner in Charleston or Fort Fisher at Federal Point?

Additionally and obviously, heavy artillery is just that: heavy. Seacoast artillery and siege pieces couldn't be haphazardly sited or they would rapidly sink into the unsupported soil. The logistics of transporting a large cannon and ammunition could be a nightmare for planners and logisticians, and even when the shipment reached its desired destination, a suitable foundation for firing needed to be selected or constructed.

This chapter outlines the different types of artillery used during the war before exploring how these weapons were individually influenced by geology. In later chapters, especially in those detailing the fighting along the coastlines, these links between geology and heavy guns will be probed further, as the soldiers on both sides of the conflict slowly began to understand the value of sand.

Types and Purpose

Artillery can be subdivided into four basic categories defined by the intended use and portability or maneuverability.[3] The easiest type of gun for transport was *mountain* artillery. These small pieces could be broken down into components and transported on horseback. *Field* artillery was more powerful and longer-ranging but more difficult to maneuver across rough terrain or through narrow passes such as those found in mountainous regions. Field artillery made up the vast majority of guns used on most battlefields from the Eastern and Western Theaters of combat. *Siege* and *garrison* artillery was more powerful, accurate, and longer-ranging than field artillery, but the pieces were also significantly heavier and more difficult to transport. *Seacoast* artillery (and naval guns) were the heaviest; their immobility led to use almost exclusively for defensive applications in coastal fortifications or along rivers, where the pieces would not need to be moved or relocated. These immobile, massive guns were often placed behind sandy or earthen parapets for the protection of the gun crew. In most cases, these guns required specialized lifting equipment for transport or even for repositioning a gun within a fortification.

Before 1861, the Federal arsenal possessed few, if any, rifled cannon. Their smoothbore guns were differentiated by the weight of their shot: a six-pounder cannon fired a (roughly) six-pound spherical projectile.[4] With the advent of the advantageous rifling of pieces, a new nomenclature was needed. Rifling meant that elongated projectiles were now common, so a single type of weapon could fire shells with a variety of different weights. For rifled guns, the diameter of the bore was commonly used as an indicator of the type of piece (e.g., a 3-inch ordnance rifle).

This terminology was simple enough to understand until Parrott rifles began to confuse the weight/diameter designations. Parrotts traditionally retained the "pounder" designation, even though the common 8-inch Parrott could fire either a 150-pound or longer 200-pound shell. The 10-inch Parrott was referred to as the 300-pounder, even though it rarely fired anything other than a 250-pound shell. It is not surprising, then, that many Parrotts carried a stamp on their muzzle with both their diameter and weight designations.

Field Artillery

HOWITZERS AND GUNS

Both howitzers and guns were smoothbore weapons commonly found as four-piece (Confederate) or six-piece (Union) field batteries.[5] Both fired a variety of solid and exploding shot over relatively short ranges. Howitzers tended to have shorter, lighter barrels that could be elevated to a higher angle when compared to a gun. These maneuverable and handy weapons used a smaller powder charge to lob exploding shells a few hundred yards across the battlefield. Guns had longer, heavier, smoothbore barrels that fired solid shot or shells with a larger powder charge to propel projectiles at a flatter trajectory. The most common field gun used during the Civil War was the bronze Model 1857 Napoleon (see figure 3.1), which fired a 12-pound shot at ranges exceeding 1,500 yards.[6] Both sides manufactured and used this type of gun in great quantities.[7] During later stages of the war, as new types of exploding shells were developed, both armies began to rebore and rifle their guns and howitzers.

FIG. 3.1. Confederate bronze Napoleon cannon on Seminary Ridge, Gettysburg. Most Confederate Napoleons had straight barrels, lacking the muzzle swell exhibited by Union guns.

RIFLED GUNS

Rifled artillery offered the same advantages over smoothbore guns and howitzers that rifle muskets did over muskets: increased effective range and accuracy. The two most common types of rifled guns were the 3-inch ordnance rifle and the Parrott. The Parrott is distinctive regardless of caliber because of the broad, thick reinforcing band that Robert Parker Parrott, the gun's inventor, added in 1860 (see figure 3.2).[8] Both rifled guns had maximum ranges approaching 2,000 yards (see table 3.1). One apparent advantage of rifled guns, with respect to sedimentary geology—their ability to penetrate parapets more than smoothbores—is negated by their poor performance with ricocheting or grazing fire: the long cylindrical bolt will more commonly dig into soft soil or strata when compared to spherical round shot.

The muzzle energies listed in table 3.1 give a very rough approximation of the striking power of solid shot when fired from each type of artillery piece. Solid shot slowed dramatically when fired over lon-

FIG. 3.2. Muddy 20-pounder Parrott rifled artillery pieces in the vicinity of Richmond, Virginia, during the summer of 1862. This detail is from a photo of the 1st New York Battery by James Gibson. Library of Congress.

ger ranges or when ricocheting, but even at longer ranges a soldier hit by shot would have felt an impact far greater than a modern rifle bullet in terms of kinetic energy. This relationship is perhaps analogous to the difference between the .45 ACP bullet fired from the venerable Colt M1911 pistol used in the First and Second World Wars and a comparable 9mm pistol bullet. The .45 ACP fires a larger, blunter bullet at a velocity only two-thirds that of the 9mm. Soldiers remarked that even though the muzzle energy of the heavier, slower bullet and the smaller, faster bullet were roughly equivalent, the .45 had a greater apparent stopping power. The solid shot was moving at a slower velocity than a musket ball or modern rifle bullet but still contained significantly more stopping or hitting power.

The Civil War fieldpiece with the highest striking power relative to its bore diameter was the imported British Whitworth gun. This breech-loading 12-pounder also had the greatest penetration depth into sediments because of its high velocity, flat trajectory, and narrow hexago-

TABLE 3.1. Field artillery specification and "hitting" power

Name	Primarily used by	Tube material	Bore (inches)	Length (inches)	Weight (pounds)	Shot/shell (pounds)	Velocity (feet/second)*	Range (yards at 5°)	Energy (foot-pounds)†
12-pounder mountain howitzer	both	bronze	4.62	33	220	8.9	~850	1,005	99,930
6-pounder gun	Both	bronze	3.67	60	884	6.1	1,439	1,523	196,299
M1857 12-pounder "Napoleon"	Both	bronze	4.62	66	1,227	12.3	1,440	1,619	396,366
12-pounder howitzer	Both	bronze	4.62	53	788	8.9	1,054	1,072	153,651
24-pounder howitzer	Both	bronze	5.82	64	1,318	18.4	1,060	1,322	321,289
10-pounder Parrott rifle	USA	Iron	~3.0	74	890	9.5	1,230	1,850	223,357
3-inch ordnance rifle	USA	Iron	3.0	69	820	9.5	1,215	1,830	217,943
14-pounder James rifle	CSA	bronze	3.80	60	875	14.0	~1,200	1,700	313,297
20-pounder Parrott rifle	USA	Iron	3.67	84	1750	20.0	1,250	1,900	485,642
12-pounder Whitworth	CSA	Iron	2.75	104	1092	12.0	1,500	2,800	419,594

* This is muzzle velocity (velocity on exiting the barrel). At range or after ricocheting, a shot or shell would be moving at a fraction of this velocity.

† Muzzle energy reflects "hitting power" of solid shot only. It is directly related to the weight of the shot and square of the velocity. Energy figures for howitzers (which typically fired exploding shells) would be overestimated. For comparison, a .50 caliber machine gun bullet has a muzzle energy of ~12,000 foot-pounds.

nal bolt. Whitworths were relatively rare on the battlefield, as finding locations to exploit their exceptional range, accuracy, and penetrative power was difficult on battlegrounds with forests or rolling terrain.[9]

Heavy Artillery

While the Union army had a distinct advantage with respect to the quantity and quality of field artillery in 1861, this was not true with heavy artillery. The seizures of undergarrisoned coastal fortifications at the start of the war by Confederate forces provided a comparatively equal number of heavy guns to the South.[10] The South supplemented this firepower by building additional pieces, albeit at a slow rate.

COLUMBIADS

Columbiad guns were the oldest weapon in the U.S. arsenal at the start of the fighting, with a design that dated from the War of 1812. This gun is considered the first piece of ordnance to be completely designed and built in the United States.[11] Compared with the original weapons from the early nineteenth century, Civil War Columbiads were larger caliber, heavier, and greatly improved in many respects (see figure 3.3). The common 8-inch variety, for example, weighed nearly 10,000 pounds and fired a 64-pound shell to a range approaching 5,000 yards. Many Columbiads were also improved during the war with reboring and rifling or by having iron-strengthening bands added to the bottom of the

FIG. 3.3. A 10-inch Columbiad positioned in Battery Semmes above Farrar's Island, overlooking the James River below Richmond, Virginia. Note how easily the gun embrasure was carved into the soft, cohesive Tertiary sediment of the Coastal Plain (and with no need for revetment). Detail of a stereographic photograph from the Library of Congress.

gun tube. Despite these enhancements, Columbiads could be dangerous to operate, with multiple incidents of exploding tubes resulting in
casualties.

HOWITZERS

Large, lower-velocity howitzers were often placed in fortifications to
protect against infantry attack. As such, they were usually located above
the land approaches to a fort, at positions separated from the main artillery embrasures of the fortification.[12] When used in siege operations,
the howitzers relied on Columbiads or rifled artillery to breach the enemy fortification's walls, at which point the howitzers' exploding shells
could be fired into the rubble and debris on the interior of the fort. The
most common heavy howitzer was the 8-inch model, which could effectively be fired at targets from 300 to 1,600 yards.[13]

HEAVY RIFLED ARTILLERY

Two similar weapons, the ~4.5 inch M1861 and M1862 rifled guns, saw
much use by both armies during the Civil War. These siege rifles fired
25 to 30 pound shells and weighed around 6,000 pounds. The Union
also used the effective James rifles in multiple calibers, including 24-,
32-, and 42-pounder varieties. Parrott rifles, with cast iron tubes reinforced with a wrought iron band, were also used in a number of different sizes (see figures 3.4 and 3.5). The 6.4-inch, 100-pounder Parrott rifle could fire a shell more than five miles, although no adequate sighting
system could assure accuracy at anything approaching this range. The
10-inch Parrott fired a 250-pound shell but was a logistical nightmare—
transporting an object that weighs more than 25,000 pounds by any
means other than ship, or perhaps rail, required special planning. Siting such a beast on most types of sediment was obviously prohibited,
with medium and coarse sand being one exception.

In general, the smaller Parrott siege guns had a lower propensity for
exploding when compared with similar but larger caliber pieces. Postwar reports by Henry Abbot for the Corps of Engineers described 4.2-
inch Parrotts that reliably fired thousands of rounds in the campaigns
around Charleston and Petersburg.[14]

The largest Parrott rifles, of 8- and 10-inch varieties, were used on or
near Morris Island, South Carolina, for the reduction of Battery Wagner and Fort Sumter, as well as the bombardment of Charleston (see
figure 3.6). The smaller variety was emplaced on the famous "Swamp

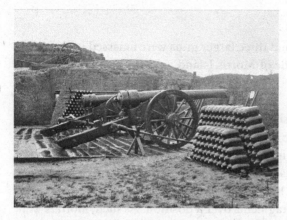

FIG. 3.4. Two 100-pounder Parrott rifled artillery pieces in Fort Putnam on Morris Island, South Carolina. A smaller fieldpiece stands guard in the background, protecting the beachfront of the fort. Detail and enhanced photograph from the Library of Congress.

FIG. 3.5. Heavy artillery in Fort Brady, Virginia, in 1864. This detail of a battery of Parrott guns was manned by Company C, 1st Connecticut Heavy Artillery. Timber and sandbags hold back the oversteepened slope of the earthen parapet. Library of Congress.

FIG. 3.6. These two 8-inch Parrott rifles were located on Morris Island and used as a breaching battery against Fort Sumter. Note the abundant sandbags used to revet the sandy parapets and gabions bracketing the gun embrasure. Library of Congress.

Angel" marsh battery, and three larger guns were amassed in the dunes toward the landward side of Morris Island.

MORTARS

Heavy mortars were designed to lob a large exploding shell into an enemy position so that the final fall of shell was nearly vertical. Using this technique made the height and thickness of a defender's fortification walls or parapets inconsequential. If properly fused, the timing of the detonation allowed the heavy shell to burst just above the enemy, raining high-velocity shell fragments over a position. As such, mortars were primarily used as antipersonnel weapons and were not considered as effective against parapets of earth or brick construction, or for destroying enemy artillery positions.[15]

Large Civil War mortars ranged in size between 10 and 13 inches in caliber and exceeded 15,000 pounds in weight (see figure 3.7). They were included in coastal fortifications and occasionally were transported by, and fired from, flatbed railroad cars.

Of all the varieties of Civil War artillery, mortars may have been the most affected by sedimentary geology. Timing fuses on mortars, especially of the Confederate variety, were highly imprecise. As a result, when mortars were trained on targets located on soft or muddy ground, their rounds often buried themselves in the sediment before detonating, rendering the impact minimal at best.[16]

FIG. 3.7. These thirteen-inch seacoast mortars were photographed near Yorktown, Virginia, in the spring of 1862. This detail of a photograph by James Gibson portrays the officers of 1st Connecticut Heavy Artillery. Their artillery position has been carved into the sandy sediments of either the Quaternary Shirley or Chuckatuck Formation. Library of Congress.

NAVAL GUNS

Naval guns were similar, if slightly larger, than siege or garrison artillery. They were used on naval vessels as well as on land. The Federal navy supplied both guns and gun crews to the land forces on multiple occasions to supplement the firepower for both sieges and defensive positions. At Vicksburg, for example, the Union navy supplied both the gun crews and 32-pounder Parrott rifled cannon for Battery Selfridge.

One advantage the U.S. Navy had over their opponents was the engineering talent of Rear Adm. John A. Dahlgren, their chief of naval ordnance. He designed modern, distinctive weapons that ranged in caliber up to 20 inches. Nevertheless, most U.S. naval artillery consisted of either 9-inch guns firing 70-pound shot or 11-inch guns firing a 127-pound round.

The Confederates also produced an effective naval gun, but on a much more restricted basis. The Tredegar Iron Works in Richmond and the Confederate foundry in Selma, Alabama, produced a Brooke rifle with sizes ranging from 6.4 inches to 11 inches in caliber. As with other unpredictable and dangerous heavy artillery from the time period, Brooke rifles had a propensity to explode. This failure has been attributed more to the quality of material and workmanship than to the design of the weapon, however.[17]

Perhaps the most effective heavy artillery employed by the southern forces during the war was foreign in design and construction. Early in the rebellion the South purchased many British guns, and some of the best ordnance in Fort Fisher and in the defenses around Mobile, Alabama, Vicksburg, Mississippi, and Charleston, South Carolina, was produced across the Atlantic (see figure 3.8). These rifled guns, manufactured by the Blakely and Armstrong firms, included the largest caliber weapons used by the Confederacy.

AMMUNITION

For longer-range antipersonnel or counterbattery fire, two types of projectiles were commonly used: shot and shell (see figures 3.9 and 3.10). Smoothbore guns fired spherical solid shot, with the weight of the projectile defining the caliber of the gun (e.g., 12-pounder). Solid shot from rifled artillery was called a bolt and also relied on kinetic energy to destroy a target.

FIG. 3.8. An Armstrong gun from Fort Fisher, North Carolina. This gun was imported into the Confederacy in 1864 and captured by Federal forces a year later. The interior of the broad landface of the fort can be seen in the distant background. Library of Congress.

FIG. 3.9. Sherman's troops remove heavy artillery shot from Fort McAllister in December 1864 after the earthen fortification was captured. Note the compacted sand the men are crossing. Library of Congress.

FIG. 3.10. Detail of a Confederate rampart outside Yorktown, Virginia, in 1862 showing various types of ammunition. An ammunition box marked "Gen. D.H. Hill, Yorktown" sits in the front center—Hill commanded the Confederate-left at Yorktown. Library of Congress.

When firing solid shot, rifled artillery had the advantage of being more accurate than smoothbore guns. Nevertheless, the spherical shot of a smoothbore cannon could still be used with devastating effect if the local geomorphology allowed. Given a slightly elevated firing position, a gently sloping landscape between gun and target, and soil that was not too soft, wet, or recently plowed, gunners could fire spherical shot to ricochet or skip across the field. Firing in this manner produced a killing swath and increased the hitting power of a shot, especially when it was fired from an enfilade, or flanking, position. If the slope was too great or the soil was too soft (or had recently been disturbed by tilling), the spherical shot tended to dig into the ground, as it would with less spherical bolts from rifled artillery.

This artillery tactic predated the Civil War. In Burn's *Questions and Answers on Artillery* (1848) he describes ricochet fire as "shot that bound or hop along the ground," which is the result of "decreasing the charge of powder, and firing the piece with an elevation ranging between 3 and 10 degrees."[18] In short, ricocheting shot was most dangerous when the trajectory of a shell could be flattened across the low-relief landscape, but not necessarily by increasing the projectile velocity.

In a summary of artillery use at the end of the war, Col. Francis Lippitt (1865) referred to low-angle ricocheting fire as "grazing fire" and praised it as "the most destructive" form of cannon fire. He further warned against muddy or wet soil because ricocheting shells in such conditions was impossible, and "shells often sink into the mud, and thus are either extinguished or explode with but little effect." Lippitt estimated that the proper use of ricochet fire increased the effectiveness of artillery by 25 to 50 percent.[19]

Sedimentary geology and differential weathering could make this force-multiplying tactic impossible, however. For example, ricocheting fire was difficult to employ on much of the battleground of Antietam: in the northern portion of the battlefield the slope was flat and nearly ideal for grazing fire because of the equal rates of weathering across the thick Conococheague Limestone.[20] Nevertheless, this portion of the battlefield was almost entirely farmland where the tilled and disturbed soil tended to absorb shot.[21] The more southern portions of the battlefield consisted of rolling terrain created by differential weathering of the limestones, dolostones, and shales.[22] In such terrain the angle of fall of the shot was too great to expect grazing unless it was fired at a tangent to the hillcrest—and in that case the shot would ricochet once then pass over the entirety of the intended target.

In his thorough review of the Battle of Gettysburg, Allen Guelzo (2013) points out that the igneous ridges of diabase provided a nearly ideal position for artillery. Cemetery Hill combines two geomorphological features that made it an exceedingly strong artillery position. First, the elevation above the surrounding softer and flatter sandstones and mudstones was within the range required for both direct and grazing fire: "Ideally, one percent of the distance to the target and never greater than 7 percent of the distance."[23] Second, and importantly, unlike most ridges of hard rock, Cemetery Hill was not so narrow at the crest that gun crews and their support teams (caissons, horses, ammunition supply) would be cramped and not have the room to efficiently operate.[24]

In other words, this ridge had a great enough elevation to provide long-range visibility, a gradient that was low enough to allow low-angle shell-fire and ricocheting fire at 600 yards' distance (beyond small arms fire), and a plateau-like ridgetop that allowed room for gun crews and their support teams to fastidiously reload.

In contrast to nonexplosive solid shot, spherical case (or shrapnel) and common shells were designed to burst in the air or on the ground in the proximity of the enemy. This was accomplished using a timed fuse that would explode the shell, spreading the enclosed musket balls (shrapnel) or fragment the shell (common).[25]

Shells offered multiple advantages over solid shot. When approaching infantry were loosely spaced or partially concealed by terrain, the exploding shells had a higher probability of producing casualties. This is true when firing at cavalry as well; hitting a horse with shrapnel will slow an advance just as well as hitting the rider. Nevertheless, inconsistent and inaccurate fuses often resulted in the shell plowing into soft sediment or soil prior to detonation, diminishing greatly the explosive and concussive effects of the projectile.

The final common type of field artillery ammunition was used at ranges of less than around 500 yards: grapeshot and canister (case). This category of projectile was essentially a tin cylinder full of round balls of different sizes that turned artillery pieces into giant long-range shotguns. This type of antipersonnel ammunition was similar to solid shot in that it became more effective when fired across hard, flat terrain where ricocheting could be expected. In emergencies and at extremely close ranges (< 150 yards) two canister rounds could be fired with a single charge, producing devastating effects on approaching infantry or cavalry.

Siege and naval artillery projectiles were even more varied than those available for field pieces.[26] As with field artillery, most rounds supplied and fired were solid shot and shell. Solid shot tended to weigh around 20 to 30 percent more than exploding shells for the same gun.

These larger guns also fired grapeshot and canister shells for antipersonnel applications. The siege and naval canister rounds were obviously much larger than those fired by field artillery (> 30 lbs. vs. < 10 lbs.), and the size of the individual balls (shrapnel) varied as well. Most field canister shells contained between twenty and forty 1- to 2-inch round balls, while the larger guns fired fewer, but much larger, balls of grapeshot. Many of the largest naval guns fired "grape" shot that was the same diameter as smaller field artillery solid shot (i.e., a 10-inch naval gun fired canisters equivalent to a barrage from a dozen 6-inch field guns firing solid shot).

Diverse Artillery, Dissimilar Applications

Whenever practicable, rifled artillery was preferred for long-range fire that required a higher degree of accuracy. However, it was slower to reload, which quality made it less desirable as the enemy infantry or cavalry approached. For use of canister, the reliable and effective M1857 Napoleon was the preferred weapon according to most post-conflict reports.[27] It was fast to reload and easy to maneuver, making it an ideal gun in close(r)-quarters fighting.

The Union army had more rifled guns during most battles than the Confederates, suggesting they should have had a firepower advantage in both quantity and range. Paddy Griffiths (1987) points out that for counterbattery duels, however, this range advantage was often diminished by terrain (and geology).[28] He provided an example from Antietam where "long range Union artillery achieved considerably less im-

FIG. 3.11. Federal Battery De Golyer on the Vicksburg Battlefield was the largest concentration of Union guns on the battlefield. This amassed firepower (22 total guns) had clear sight-lines towards the center of the Confederate works.

portant results than had been hoped. . . . The Confederates drew their cannon into the folds in the ground." These folds are represented by the swales created as limestone weathers more quickly than dolostone; almost certainly the Confederate guns were hidden by the rolling terrain of the Elbrook Formation.

One general trend was true in the tactical use of Union and Confederate artillery in both theaters of war: Federal batteries were both larger (six guns vs. four) and often more concentrated (see figure 3.11). Confederate batteries tended to be more spread across the battlefield, offering more flexibility but lacking the hitting power against one portion of the Federal line.

While sedimentary geology often diminished the impact of artillery during a battle or siege, the same combination of sediments could prove highly effective at protecting a gun and crew during combat. The next chapter explores how sediments were exploited on the defensive: piled for protection.

CHAPTER 4

Geology and Protection

Fortifications

To Make Strong

Fortifications, from the Latin *fortis* ("strong") and *facere* ("to make"), are constructed because they represent a force multiplier. This is true of massive brick fortifications constructed along the coast or simple breastworks thrown up in anticipation of an approaching fight. Napoleon's quotation "With mediocre troops one must shift much soil" was especially valid with Civil War soldiers.[1] While Union and Confederate infantry did not suffer from a lack of courage, they were, for the most part, highly inexperienced. In many instances troops entered combat with less than a month of drilling—the very definition of a "citizen-soldier."[2] In combat, any type of protection, whether a parapet, rifle pit, split-rail fence, linear trench, or preexisting stonewall, would offer some perception of protection, allowing a soldier an increased sense of security. More confident soldiers performed better when fighting and were more likely to deliver accurate rifle fire. They were also more likely to follow the command to counterattack, if the circumstances dictated. The length of time required for reloading single-shot muzzle-loading muskets—and the need to stand while doing so—only increased the sensation of vulnerability and amplified the value of cover and protection. Concealed and protected soldiers were also, obviously, harder to maim and kill by an attacking enemy.

The famous early nineteenth-century Prussian military theorist Carl von Clausewitz explained in his writings how proper entrenchments could both enhance the defensive effectiveness of troops and, at the same time, improve their (subsequent) offensive capabilities.[3] He advocated for allowing soldiers to resist the enemy's advances while in defensive works until the other army was weakened, at which point offensive tactics would have a greater probability of success.

Fortifications improved morale as well as defensive firepower. It is surprising, then, to find that breastworks and entrenchments were not consistently constructed during many of the battles through the first half of the war. This lack of field fortifying might be explained by a number of factors. In some battles troop movement was too dynamic to allow time for the construction of elaborate breastworks or parapets. On some battlefields local geology and thin soils prohibited digging. Some commanding officers felt the construction of fieldworks might diminish the offensive aggressiveness of their soldiers, perhaps causing a reluctance on the part of the men to leave their fortified positions when ordered to attack, with a resulting deadly delay in coordination. Note that the celebrated rallying quote by Confederate brigadier general Barnard Bee at First Manassas about Thomas J. Jackson was "There stands Jackson *like* a stone wall," not "There stands Jackson *behind* a stone wall."[4]

Speaking of stone walls, soldiers and commanders in both armies were quick to take advantage of preexisting anthropogenic or natural breastworks whenever available. In 1862 alone, the Confederates made effective use of the famous Stone Wall at Fredericksburg, the Sunken Road at Sharpsburg, and the unfinished railroad grade at Manassas; the Federals repeated this tactic by exploiting the limestone cutters (karrens) at Stones River.

Occasionally, these defensive positions were made even more formidable because they were invisible to the attacker. A large portion of the Stone Wall at Fredericksburg had sod-covered earth piled against it, causing it to blend into the slope of Marye's Heights and disappear. At Antietam, limestone outcrops provided similar cover for other Confederates. Col. Ezra A. Carman of the Thirteenth New Jersey described this effect when hidden rebels near the West Woods fired upon his men: "The men were being shot by a foe they could not see, so perfectly did the ledge protect them."[5]

These defensive positions were made even more formidable by the development of two important innovations in weapon's technology. First, the introduction of rifle muskets firing expanding lead minié balls extended the defensive range in front of an improved field position, given adequate sight lines. Second, the introduction of the percussion cap to replace the flintlock mechanism allowed a defensive position to remain at strength even in rainy weather.

As the war progressed, entrenchments became more ubiquitous, and by the end of the war some battles began to resemble the stalemated fighting in France during the First World War. As expected, offensive tactics changed with the increased use of fortifications. Night attacks, while difficult to coordinate, execute, and exploit, became more common in Grant's Overland Campaign to capture Petersburg and Richmond and during fighting on the Coastal Plain and barrier islands. Commanding officers were (slowly) learning that headlong attacks on well-fortified positions were hopeless, and experimental and innovative approaches, including tunneling, began to be more commonly employed.

Sediments and Temporary Field Fortifications

By far the most important military engineer, teacher, and theorist in the United States prior to the Civil War was Dennis Hart Mahan, who both lectured at the U.S. Military Academy and wrote the manuals used by the students, professors, and professional engineers of the army.[6] In his classic 1856 *Treatise on Field Fortifications*, Mahan applied the term "intrenchment" to all field fortifications.[7] He stated plainly that all intrenchments or fieldworks had three general purposes. First, all works must provide defensive cover. Second, they should provide improved firing positions. Finally, they should prove a hindrance for the attacking force or "present an obstacle to the enemy's progress."[8] The simplest way to meet all three criteria was to dig a ditch and use the excavated sediment to construct an embankment, or parapet. The ditch provides an obstacle, and the parapet provides both an obstacle and protection and, if constructed carefully, perhaps even an enhanced or raised firing position.[9]

Mahan thought of his intrenchments as both a force multiplier and a defensive measure that could be critical for offensive tactics as well. Sheltered defending troops could provide a decisive counterattack after a failed assault against their works, striking the retreating enemy when their command structure was in disarray.

Mahan went into great detail describing and illustrating the proper cross-section of a parapet. Every aspect of construction was influenced by the building material and, as a result, sedimentology. In this section of the text, the basic parts of the parapet—as differentiated by Mahan—are described and illustrated (see figures 4.1 and 4.2). Next, each sec-

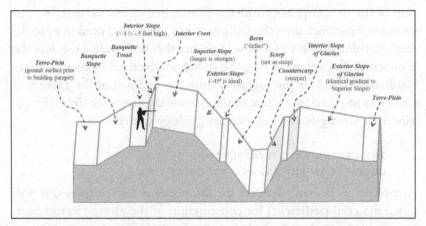

FIG. 4.1. Basic block-diagram of Mahan's parapet illustrating the proper dimensions and design for the earthen wall.

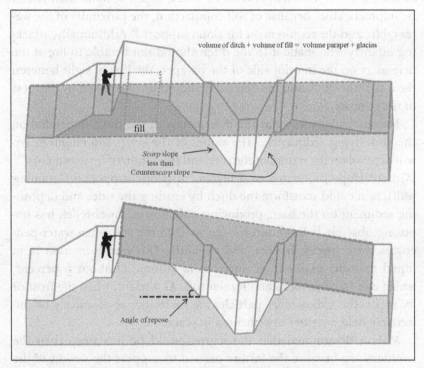

FIG. 4.2. Mahan determined that the volume of the parapet was a function of how much fill (e.g., rocks, timber, split-rail fencing) was laid down prior to burial by the sediments from the ditch. Matching slopes (darker gray) were critical for providing the most effective sight- and firing-lines.

tion of the fieldwork is considered from a geological perspective. For example, how thick does the sedimentary parapet wall need to be to defeat incoming artillery? Or how steep can the ditch walls be before the sediment collapses, rendering the trench ineffective? Numerous high-resolution Civil War photographs from the archives of the Library of Congress are used to document the relationship between defensive engineering strategies and sedimentary geology as well.

THE PROPER WAY TO PRESENT AN
OBSTACLE TO THE ENEMY'S PROGRESS

According to Mahan's *Treatise*, the ditch served two purposes: it provided material (sediment) for construction of the above-ground parapet, and it created a physical challenge and obstacle for attacking infantry. The ditch had a minimum depth of 6 feet, and the width was not to exceed 20 feet. Mahan also noted that a depth of more than 12 feet is "impracticable" because of soil compaction, the proximity of the water table, and the requirement for slope support.[10] Additionally, attacking infantry who made it to the ditch should not be able to fire at the defenders on the interior side of the parapet; the line of sight between the top of the ditch and the defenders should be obstructed by the crest of the parapet.

Mahan understood that the dimensions of the ditch depended on the underlying sediments: "The slopes of the scarp and counterscarp will depend on the nature of the soil, and the action of frost and rain."[11] Without slope protection or a revetment (slope support or retaining wall), rain could transform the ditch by eroding the sides and depositing sediment on the base, producing a shallower, lower-relief, less imposing obstacle.[12] Frost and ice can weaken the soil when water penetrates pore spaces, freezes, and expands. This transformation from liquid to solid results in an increase in volume of about 9 percent, which can fracture rocks and loosen soil. As a result, when the frost or ice melts, the sedimentary particles are no longer as cohesively or collectively held together or cemented in place.

Mahan determined that the scarp should be less steep than the counterscarp because the former needed to support the weight of the parapet. The weight of the parapet could cause the top of the scarp to become an oversteepened slope when the earth on the fringe of the parapet was forced horizontally.[13]

Drainage of the ditch was not a critical concern for engineers.[14] If the base and slopes of the ditch are composed of a sediment with a high clay content, infiltration will be prevented because of the sediment's poor permeability. With a large quantity of rainfall, a shallow moat might be produced, and with even a moderate amount of precipitation the base of the ditch would be transformed into a muddy quagmire. Either way, the obstacle is enhanced.[15]

According to Mahan, excavation teams needing to fortify a position should work in groups of six or seven men: one or two men to loosen the soil with a pick, two men to shovel on the scarp, two more to shovel on the counterscarp, one man to spread the loose sediment on the parapet, and one more to ram or compress the repositioned dirt. In soils that were not excessively rocky or clay-rich (cohesive), Mahan estimated that each man should be able to dig out around 6 cubic yards of sediment per day.[16] Experienced troops working in sandy soil could probably excavate more than 10 cubic yards in a day. All of this assumed, of course, that the men were supplied with excavation tools like picks and shovels, which was usually not the case during the war.

THE PROPER WAY TO INTERCEPT THE ENEMY'S MISSILES

The most critical portion of any defensive fieldwork is the parapet. A well-constructed parapet was an obstacle to attacking infantry and enhanced the effectiveness of the ditch from which it was constructed. Additionally, the parapet provided an elevated firing platform and shield from enemy "missiles," as Mahan termed musket and minié balls and artillery shells.[17] Sedimentology is especially important with parapets because the type of sediment (grain size) determines both the required thickness to defeat incoming gunfire and the degree of reinforcement and support required to keep the parapet walls from collapsing. Parapets constructed of sand or especially sandy soils required less labor to excavate and pile but more work to reinforce and maintain. Some parapets built in clay-rich regions required few, if any, reinforcing or retaining walls. These parapets would require more effort to construct but could still be fabricated quickly because there was no need for extensive use of lumber, sandbags, or sod coverings.

Mahan divided the parapet cross-section into several connected and precisely planned slopes (figure 4.1).[18] The portion of the parapet that was closest to the ditch was called the *berm*. Mahan considered the berm a necessary deficiency in the fortification: it was required to support the parapet and to prevent the scarp from collapsing, but it also provided attacking infantry with a foothold to scale the outer face of the parapet. The berm might even provide a foundation for ladders if the parapet had a particularly high-angle slope face. The width of the berm, Mahan decided, was entirely dependent on the sediment: "In firm soils, the berm may be only from 18-inches to two feet wide; in other cases, as with marshy soils, it may require a width of six feet."[19] "Firm soil," as described by Mahan, was roughly a fifty-fifty mixture of sand and clay, while "marshy soil" had more clay and silt, as well as more moisture.

The *exterior slope* (figure 4.1) of the parapet is the key face that will absorb the majority of the incoming artillery rounds. Mahan specified that the elevation and width of this face were the same length, thus the face would be angled at 45° facing the enemy. This slope would be difficult to scale, and theoretically, any sediment thrown vertically by artillery strikes would fall back to the same approximate location, leaving the parapet only slightly diminished in efficacy.[20] From a geological perspective, this 45° angle also represents the critical *angle of repose*—the greatest angle of dip at which a sediment can be piled before it begins to slump. Any angle exceeding the angle of repose will require a revetment of some type to stabilize the wall. In general, rough and angular sand can be piled at a higher angle than smooth and round grains.[21] The addition of cohesive clay increases the angle, as does the addition of a small amount of water.[22]

Mahan's mixture of soil and clay (earth) would have an angle of repose of 35°–45°, which he recommended as the default parapet slope for earthworks. Slightly wet sand is similar. Dry sand, in contrast, has an angle at least 10° lower, meaning that coastal fortifications built with exterior slopes of 45° would need lumber, sandbags, or sod reinforcement (revetment) unless they could be kept moist, which was impossible. For most coastal sand forts, the angle of repose for the walls would be even lower because of the source of the sand; the closest abundant source of sediment would be the beach or dunes, both localities providing mature sediments that would be well sorted and have rounded and spherical grains.[23]

The top of the parapet is referred to as the *superior slope*. The length of this slope was determined by the thickness of parapet needed to defeat the expected artillery of the enemy. The slope was calculated from a combination of three factors. First, the slope downward from the crest where the defending infantry was positioned should allow an adequate field of fire across, but not into, the ditch (figure 4.2). If a soldier could fire into the ditch, it was reasoned, then an enemy sheltering in the ditch could effectively return fire. Second, the slope should be angled to allow incoming rounds to be deflected above the defending infantry and artillery positioned behind the parapet. Third, the slope should not be so great that the crest is easily penetrated because it is too high and thin.

One note of caution was made in the *Treatise* regarding the type of sediment used to form the superior slope: if the sediment was poorly sorted or contained excessive amounts of pebbles and cobbles (gravel), soldiers were instructed to cover the surface with vegetation of some type.[24] Otherwise, incoming artillery or small arms fire might create secondary projectiles when they hit the coarse sediment.

The interior wall of the parapet was divided into three sections: two slopes separated by a nearly horizontal standing platform for the defending infantry called the *banquette terrace* (figure 4.1). Soldiers would stand behind the interior slope and fire over the crest of the parapet at the enemy. The interior slope was almost always an oversteepened slope requiring revetment because soldiers preferred to have the elevated portion of this wall as close as possible to their heads and torsos for protection; an angle closer to that of repose would have left them feeling vulnerable and exposed (see figure 4.3).[25]

The height of the interior slope was a compromise between the amount of sediment available for construction for the parapet and the elevation needed to provide cover for a standing soldier as he fired and reloaded his rifle. Before breech-loading rifles became widespread, long guns loaded from the muzzle end. This required a three-foot-long ramrod to be extended above the end of the barrel to push the powder charge and bullet down the barrel. These tasks are highly awkward when sitting or lying on the ground, so protection of a soldier while standing was a necessity to ensure rapid shooting and maximum small arms firepower.

If the height of the interior slope was more than five feet, some soldiers might not be able to fully depress their weapons along the superior slope, creating a vulnerable dead zone in front of the works when the enemy drew closer.[26] If the slope elevation was too low, of course, a larger target was exposed to the enemy, including sharpshooters. For extra protection soldiers often added a slightly elevated log to the top

FIG. 4.3. Timber revetments used in this enlarged and enhanced photograph of the Federal line near Fort Morton, Petersburg Virginia. Note the higher angle of repose for the non-revetted clay-rich sediments. Library of Congress.

of the interior crest. This "headlog" provided extra protection when reloading and allowed the soldiers to fire through the gap created between the upraised log and the crest of the parapet.

The width of the banquette terrace was determined by the number of defenders intended to be positioned along the interior of the parapet (one rank or multiple). At a minimum, the terrace needed to be wide enough to allow a soldier to rest the butt of his musket on the ground as he reloaded while another soldier passed behind him. A well-constructed banquette terrace would be slightly sloped toward the interior of the fortification to allow for the proper drainage of precipitation.

The final interior slope was called the *banquette slope*. This surface was shaped below the angle of repose and supported the weight of the banquette terrace. A lower angle slope provided more support but took up valuable flat interior space behind the parapet. A lower slope was also easier for supporting members of the garrison to climb to provide ammunition or reinforcements.

Different Sediments, Different Revetments

Freshly dug and piled earth is particularly prone to erosion, and vulnerability to this threat escalates as the proportion of sand increases and the amount of clay decreases. On a barrier island, an unreinforced sand parapet would not last a month—wind, precipitation, and waves would level it quickly. Without proper support, a newly finished earthwork will begin slumping and suffering from various forms of mass wasting almost immediately. Any slope over the angle of repose for the sediment or earth mixture needs reinforcing support as it is constructed.

Mahan recommended the use of multiple types of revetments for this required structural support. A revetment, he instructed, "consists of a facing of stone, wood, sods, or any other material to sustain an embankment, when it receives a slope steeper than the natural slope."[27] The choice of materials was dependent on how long the parapet was intended to be used and what materials were locally available for construction.[28] A sand fort on a barrier island, for example, would not have had a supply of stone, and valuable timber and lumber were scarce, so the obvious choice was sod from the fringes of the back-barrier marsh.[29]

Of all revetment types used during the Civil War, one has the longest legacy: sandbags. Mahan preferred vegetation-derived materials (wood planks, timber, sod, wicker structures called gabions and fascines) over sandbags because he worried about the maintenance demands created by deteriorating canvas. Nevertheless, empty bags were easy to transport, they could be filled quickly with the abundant sand found on or near most battlefields, and sandbags were effective protection against both artillery and small arms. One additional advantage of sandbags over timber or stone is that sandbags do not splinter after a high-velocity impact—no secondary projectiles.

Another common revetment material that required sediment to be effective was the gabion. A gabion was constructed of sticks in a basketform with flexible half-inch twigs used to tie them together. These cylindrical wicker drums were around two to three feet in diameter and three feet high. When filled with sediment they could stop a bullet or offer support for an overly steep slope. Often, for example, the interior slope of a parapet was lined with gabions (see figures 4.4 and 4.5).

One major drawback to using gabions occurred when fine dune or beach sand was used as fill. On Morris Island, for example, gabions

FIG. 4.4. Sand-filled gabioned parapet and traverses of "Fort Hell" (Fort Sedgwick), Petersburg, Virginia in the spring of 1865. Note empty gabion lying on its side on the right. Library of Congress.

FIG. 4.5. A second photograph detailing the use of gabions as revetments on the interior of Fort Sumter after the fortification had been reduced by Union artillery late in the war. These gabions were filled with both sand and brick rubble. Library of Congress.

were replaced by Federal sappers with sandbags when the fine sediment was found to leak from between the interlocking sticks. Another twig-based revetment material, the fascine, did not require sediment. These bundles of straight sticks could be used with gabions or stand alone to hold back a potentially slumping slope. Fascines were often used near the muzzles of larger artillery pieces in place of sandbags, because the cloth of the latter revetment material tended to rupture or explode from the shock waves created by the gun blast.

FIG. 4.6. Sod revetments on the interior of Fort Putnam (previously Confederate Battery Gregg) on the northern tip of Morris Island, South Carolina. Library of Congress.

When time allowed or fortifications were intended to be more than just temporary in nature, Mahan suggested lining the parapet with sod (see figure 4.6). He was very specific with how sod should be cut and how it was to be formed on the exterior of the parapet. The sod needed grass "of a fine short blade" with "thickly matted roots."[30] Long grass should be mowed and cut into two sizes of 4.5-inch thick blocks—either 12 inches by 12 inches or 12 inches by 18 inches. Two layers of alternating blocks were to be laid on the parapet so that their edges overlapped because of the different sized blocks.[31] The first layer was laid with the grass side down, toward the parapet. The second layer was placed on top of the first so that the root side was down and the grass side faced upward, toward the sun. Placement of the sod blocks in this manner left the root sides together, and the entire revetment could be compressed onto the parapet with rams or, if desired, fastened with wooden spikes.

Perhaps foreseeing the problems with a sod cut from moist, clay-rich soils (e.g., those found in a back-barrier island marsh), Mahan instructed that wet sod soil should be partially dried before use. Clay is porous but relatively impermeable. This sediment can hold much water, but the pore spaces are not well connected, prohibiting water from flowing through the soil easily. When clay dries it loses a substantial portion of its volume, and a sod revetment constructed of wet, clay-rich blocks would shrink, leaving gaps in protection of a few inches between blocks. This is why mud cracks are such a common sedimentary structure in mudstones and shales.[32]

This manner of sodding worked best along the northern and mid-Atlantic coastlines where rich, thick grasses and reeds were plentiful. On many marshes from the southern barrier islands, small tufts of Bermuda grass were carefully planted across the expanse of the parapet. To encourage their growth, a layer of soil or garden loam was added, and they were regularly watered.

Defensive Accoutrement

When time allowed, multiple other defensive enhancements were added to the fieldworks to slow the approach of attacking infantry or cause a delay in their ranks once the enemy came within rifle range. Some of the sediment dug from the ditch could be thrown to the opposite side of the depression from the parapet, creating a gentle slope for attackers to climb as they approached the fieldwork. This shallow slope was called a *glacias* (figure 4.1), and the degree of slope was matched to that of the superior slope (figure 4.2). If the fieldwork were engineered properly, the enemy's final approach to the fort was uphill and directly into the firing plane of the defenders.

Felled trees were often also used for multiple defensive purposes. Cutting trees in front of a parapet was a necessity to reduce cover for attacking enemy infantry and increase sight lines and defensive fields of fire. These trees could be cut and felled in a manner to provide several types of obstacles. If the tree trunk was straight and around a foot and a half in diameter, it could be cut into ~10-foot lengths to form part of a panel used for palisading. The *palisade* will form a vertical or slightly tilted wall slanting toward the enemy. For stability, the wall was buried in a narrow trench to a depth of approximately 3 feet. For less cohesive, sand-rich sediments this depth might be greater, and the tree trunk will need to be slightly longer. The top of the tree trunk was sharpened, producing a 7-foot-high timber wall with a pointed, piercing peak, and loopholes could be cut between logs to provide protected firing positions for soldiers.

Mahan suggested constructing the palisade in the ditch at the base of the counterscarp. In this location the palisade created an imposing obstacle without diminishing the visibility of the defenders or restricting their fields of fire. It was also difficult for an attacker to seek pro-

FIG. 4.7. Fort Putnam (formerly Battery Gregg) was surrounded by a fraise. Detail (below) appears to show an abatis has been added behind the fraise. Sullivan's Island, across Charleston harbor, is visible in the background. Library of Congress.

tection behind the palisade while on the steep counterscarp. The positioning of a palisade in the ditch also protected it from the enemy's artillery—the greatest threat to a log wall.

When horizontal, or nearly so, a palisade was called a *fraise* (see figure 4.7). Added to the berm of a parapet, a fraise created an even more difficult obstacle to overcome while at the same time remaining hidden from artillery and not diminishing the defender's field of view. The logs in the fraise did not need to be directly adjacent to each other to be effective; they only needed to be close enough together to prevent an enemy infantryman from squeezing between the trunks. The gaps between the fraise elements represented areas of exposure for the attacker, providing firing corridors where the enemy could be cut down while attempting to circumvent the obstacle.

Smaller portions of the tree, not used in the palisade, could be used

FIG. 4.8. This linked chevaux-de-frise provided protection for Marietta Street in Atlanta, Georgia. Note the presence of the photographer's wagon and tent darkroom in the background. Library of Congress.

FIG. 4.9. Detail from a stereographic image showing massed abatis in front of the ditch along the Union lines at Petersburg, Virginia. Note the soldiers (left) for scale. Library of Congress.

to make either *cheval-de-frise* or *abatis* (see figures 4.8 and 4.9). Chevaux-de-frise were originally constructed as a defensive anti-cavalry measure in the Friesland region of Germany (and the Netherlands) a century before the American Civil War. The body of this obstacle was a 9–10-foot-long round or square piece of lumber perforated at right angles with sharpened spikes. Each cheval-de-frise could be bolted or chained with the others to produce a nasty deterrent to attack for infantry or men on horseback.

While palisades used the trunk of larger trees and chevaux-de-frise used smaller trunks and branches, abatis used nearly the entire tree. Cut trees and branches of all sizes were positioned so their tops faced perpendicularly away from the defensive fortification lines or entrenchments. The branches were trimmed and sharpened and interlocked with supporting trees to form a simple but effective barrier.

The creation and placement of abatis provided two benefits to the defenders. Cutting trees improved sight lines from within the fortification, increasing the effective range of small arms and artillery. It also made gun laying easier, and the felled trees provided an obstacle that would slow an enemy's attack. This delay provided additional time to discharge and reload the slow-firing weapons of the time. This time-consuming hindrance was especially valuable to defenders who were wary of a possible night attack. In some cases, the trees could be felled so they were not completely cut through and separated from the stump—the remaining uncut portion of the trunk anchored the tree to the ground and inhibited attackers from quickly lifting or displacing the obstacle.

There is an indirect but important relationship between these wooden obstacles and sedimentary geology. The species of tree locally available for modification and defensive utilization was dependent on the climate and soil, and the soil was directly influenced by the parent rock and type of clastic particles present, as well as the local geomorphology.[33] Hardwoods in the Mid-Atlantic could produce abatis, chevaux-de-frise, or palisades, while the sandier soil from the Coastal Plain of North and South Carolina and Georgia provided pine trees with perpendicular branches that would have a limited effectiveness when used as abatis. Saltwater-tolerant palmetto trees found on the southeastern Atlantic and Gulf coasts were used extensively on the palisades of coastal sand forts. They were particularly effective in defensive works because the soft and flexible wood was known to resist splintering when hit with rifle fire or solid shot.[34]

Sedimentary Determinism and
Fortification Construction

The choice of field fortification is dependent on multiple interrelated parameters. The first of these was the amount of labor that was available to an engineer tasked with construction and the amount of time before the site would be menaced by the enemy. Also of critical concern were the length of time the defensive work was intended to be occupied and the probable nature of any future attacking force—infantry, artillery, cavalry, naval bombardment, or perhaps combined operations. Of course, the anticipated strength of an enemy's assault would need to be accurately anticipated as well.

Next, it was important to determine what construction supplies were available: What were the geology and geomorphology of the region, and what soil or sediment type would be used for construction? How thick was the regolith,[35] and how deep was the bedrock? Additionally, what species, quantity, and maturity of trees were locally available? Finally, the availability and quantity of excavation tools needed to be considered. The value of a shovel continued to grow during the war as entrenching and parapet construction became commonplace.

By early 1864, rapid digging of earthworks was the norm for both armies in the Eastern Theater of conflict. Entrenching was even more frequent during the Atlanta Campaign in the west. When immediate temporary defensive works were needed, the first choice was whether the ditch should be on the interior or exterior of the parapet. If the ditch was behind the parapet (interior), away from the enemy, the defending soldiers could stand in the depression, and the parapet wouldn't need to be as high to offer protection. A lower interior-ditched parapet is also a smaller target for artillery. This was a lesson quickly learned fifty years later across the Atlantic.

If the ditch was dug outside the parapet (exterior), it will serve as an obstacle and defenders would be able to stand at an elevated, but still protected, position. A slightly elevated position is also better for defensive artillery, especially when firing exploding shells or ricocheting solid shot, grapeshot, and canister.

Whether interior or exterior, both sets of ditches and their corresponding parapet were effectively defensive enhancements that could be constructed overnight or, if needed, in a few hours. One common method of expediting construction was to add fill to the base of the parapet, laying down fencing, trees, or other debris prior to piling the earth. Debris added volume to the parapet without substantially reducing the structure's resistance to penetration, so there was no tangible penalty for the reduced expenditure of labor.

When time allowed or when semipermanent fortifications and siege works were prepared, parapets could be built higher and revetted. With even more available time for construction, obstacles could be added and interior works like bombproofs and traverses could be fabricated to obstruct enfilading or indirect fire.

Of course, commanding officers often had time to order the construction of fortifications, yet they demurred. The noted scholar on the employment of earthworks during the Civil War, Earl J. Hess, explained this reasoning during Lee's Maryland Campaign: "Antietam fits nicely into the trend many Civil War commanders displayed. They were eager to dig in during the first half of the war only when they occupied rough terrain that in itself inhibited tactical movement, but in open ground they did not want to tie their forces down unnecessarily to the defensive."[36]

On other battlefields, ditches and parapets could not be constructed because of the shallow depth of bedrock. On the Stones River battlefield, only half the initial Confederate line was entrenched despite the Confederates occupying the position long before the Union army arrived near Murfreesboro. The shallow limestone bedrock made digging impossible under most of the wooded terrain and in the cedar glades. The most famous defensive position where geology prohibited digging was the center of the Federal "fish-hook" line at Gettysburg. The diabase underlying much of Cemetery Ridge is found under a very thin regolith, and in some locations around the Copse of Trees the bedrock is exposed at the surface. Nevertheless, there are portions of Cemetery Ridge that could have been entrenched but were not.[37]

Sandcastles

The durability of sand parapets and bombproofs was well witnessed at Fort Fisher and Battery Wagner (see part V: To Take the Coasts). No building material offered the combination of ease of excavation, resistance to impact and penetration, and simplicity and rapidity of repair as sand. Nevertheless, the fine-grained quartz particles offered countless challenges to engineers and infantry garrisons.

FIG. 4.10. Two-gun embrasure containing 8-inch Parrot rifles revetted extensively with sandbags. These guns were positioned to fire into the southern- and eastern-facing walls of Fort Sumter. Note the use of gabions around the embrasure openings instead of less shock-resistant sandbags. Library of Congress.

Aeolian (wind-driven) transport of sand would erode a dry parapet quickly if it wasn't well sodden. Ocean waves and tides, as well as storms, were a constant erosional threat. For a coastal fort wanting to ricochet artillery fire across the water at approaching enemy ships, the ideal location was found as close as possible to the high-tide line. Thus, the best location to maximize firepower was at the lowest elevation on the island and the most vulnerable with respect to beach drift and storm erosion. Often planners had little choice with location; Battery Wagner, for example, essentially stretched perpendicularly across the island from the back-barrier high marsh to the high-tide line on Morris Island.[38]

Sand's granular nature limits penetration from shells and at the same time allows individual grains to penetrate other materials and structures. Soldiers and engineers alike were frustrated to watch fine barrier-island sand pour through gabions or revetment planking.[39] The result of this leakage was a gradual increase in the use of sandbags, and the quantity of finely woven canvas sandbags used during later coastal campaigns was truly impressive. The Union infantry used an estimated 46,175 sandbags on one South Carolina barrier island alone (see figure 4.10).[40]

Canvas bags were not as durable as the modern synthetic equivalent, and the result was a never-ending maintenance requirement for fortifications in the harsh conditions of the coastal southeastern United

States. Engineers on the islands around Charleston estimated bags would last around four months before they needed to be replaced.[41]

Fine sand with little clay is also a material easily carried by wind. Consequently, the sea breezes of the barrier islands deposited sand inside or on top of anything that was not covered and protected. This, unfortunately, included revolvers, rifles, and cannon. Quartz sand is also a particularly hard common mineral, and quartz tends to break in a conchoidal manner—smooth faces with very sharp edges.[42] As a result, sand can excessively wear the bore of a rifle or cannon or prevent firing entirely when it inevitably penetrates a firing mechanism.[43] And, unlike soldiers from later wars who could fieldstrip and clean their weapons at any time, Civil War soldiers had no way to repair a pistol or rifle fouled from sand. Even dropping a rifle on the beach or in the dunes could render it inoperable.

Artillery was no less susceptible to fine sediment, and gun crews had to be cautious of sand blown into a gun position or thrown up into the air by the concussion and blast of firing. One major drawback found with sandbags in artillery embrasures was their tendency to rupture from the concussion of large-caliber weapons. As such, they often had to be replaced in many positions with gabions, fascines, timber, or wood planking.

The challenging nature of sand as a building material will be revisited in more detail later in the book when the struggles over the beaches of North and South Carolina are discussed. Before doing so, however, the role of artillery and earthworks are discussed for two other physiographic provinces underlain by ancient, hard sedimentary rock: the Valley and Ridge and the Piedmont.

PART II

Hard Rocks and High Ground
The Piedmont and Valley and Ridge

We can fight Rebels but not mud.
> —Elisha H. Rhodes, Second Rhode Island Infantry

Boys, give 'em the rocks!
> —Infantryman of the Stonewall Brigade,
> after running out of ammunition

CHAPTER 5

Durable Rocks and Defensive Stands

Sedimentology and Scale:
Grand Military Strategy

Robert E. Lee launched two invasions into Northern territory, one in 1862 and a second a year later. Both efforts exploited sedimentary geology in an identical manner: marching on sedimentary (carbonate) rock and using ridges of harder metamorphosed sedimentary rocks to shield and protect the army, all while exploiting the hard-rock mountain passes at every opportunity. The Federal military command, in contrast, often failed to follow these same tenets. At Second Manassas, for example, Lee needed to march Longstreet's Corps as rapidly as possible across the Blue Ridge province to join Jackson's men on the Triassic sandstones and shales of the Piedmont. To do so, the First Corps would need to cross through Bull Run Mountain and the potential bottleneck of Thoroughfare Gap. Had Federal infantry, or even dismounted cavalry, been positioned in large number in this pass through the mountains, Longstreet would have been delayed and John Pope and the Union Army of Virginia might have crushed Jackson's isolated Second Corps. Instead, the quartzite ridges were left only lightly protected, and Longstreet was able to push through and join Jackson on the rift-valley sediments west of Bull Run.[1]

During his invasion of Maryland a month later, Lee would exploit the mountain gaps to save his army in a different manner. In early September Lee found his divisions spread all across the carbonate rocks of the Great Valley. When George McClellan unexpectedly began to move quickly (for him) from Frederick with his much larger force, Lee suddenly realized how vulnerable this dispersion of his troops left his own army. To remedy the situation and buy the time necessary to reassemble his forces, Lee sent a portion of Longstreet's Corps under the direction of D. H. Hill and Lafayette McClaws to hold the quartzite-lined gaps in South Mountain. This they did, but only for a single day before giving way to the strong Federal assaults. Nevertheless, Lee's gambit provided him with enough time to assemble the majority of the Army of Northern Virginia around Sharpsburg.

The invasion of Pennsylvania in June 1863 followed a similar geological path. Lee marched north on soft, gently undulating, easily erodible carbonate sedimentary rocks in the Hagerstown and Cumberland Valleys. Hooker's, and later Meade's, larger Army of the Potomac remained on the other side of the quartzite-rich South Mountain. Lee first used the harder sedimentary and metamorphosed sedimentary rocks to screen his movements, before rapidly passing through the hard-rock ridges to assemble his army north and west of Gettysburg.[2]

For both of these invasions, the Army of Northern Virginia was living off the land as it marched. The soils over the limestone and dolostone of the Great Valley are fertile, as are those above the Triassic sandstones, siltstones, and shales around Manassas and Gettysburg.[3] In the Western Theater, Grant would find it possible for his Army of the Tennessee to sustain itself during the Vicksburg Campaign when he ventured away from his supply lines onto the Cretaceous sediments of central Mississippi. Such invasions of foreign territory would have been impossible in regions where local agriculture failed to support tens of thousands of men on the march.

Typically, about half of a soil is composed of sand, silt, and clay, and the mineral composition of the underlying bedrock and maturity of the soil are critical for producing a topsoil that is good for farming. The Armies of Northern Virginia and Tennessee found it possible to operate without an extended and extensive supply line on the Piedmont and Valley and Ridge of Maryland and Pennsylvania and the Coastal Plain of Mississippi; such unsupported maneuvering in the Appalachian Mountains or the wetlands of the Mississippi River Valley would have been precluded by the local soil type.

By the end of 1863 such independent operations became less inviting; the North controlled the Mississippi River, so shipborne transport was feasible throughout most of the Western and Trans-Mississippi

Theaters. In the East, the agricultural landscape of much of Maryland and Virginia had been devastated by two years of foraging, raiding, and burning. Any movement by an army group greater than a division would need a long line of supply—unless operating in the previously unscathed Deep South—and the source location of the arms and munitions, as well as uniforms and shoes, was also, in part, related to sedimentary geology.

The Shape of the Land: Geomorphology

Geomorphology is the study of landforms and topographic features on the earth's surface. The geomorphology of a battlefield is dependent on three primary (and interrelated) factors: geology, climatology, and surface processes. The geology of a battleground will be expressed on the surface because of variations in the hardness and tilt of the strata: hard rock, for example, weathers more slowly than soft rock, resulting in higher elevations. Where battlefields are underlain by a variety of rocks with different degrees of durability, the differential weathering between rock types will result in high ground that is preferential for artillery locations or defensive troop positions. At Gettysburg, for example, Little Round Top and Culp's Hill are underlain by a hard and durable igneous rock (diabase), which is more erosion resistant than the surrounding sedimentary sandstones and siltstones. It is not coincidental, then, that George Meade elected to anchor the right and left flanks of his "fishhook" line on harder rock. This rugged, elevated terrain was easier to defend compared with the softer rocks his army was forced to relinquish on the first day of the battle.

The landscape and rock strata are altered by surficial geological processes. On the largest scale these processes include compression and uplift or extension and subsidence of entire regions. Erosion and deposition from precipitation and runoff, rivers, wind, and even glaciers can further modify the surface of the earth.[4] These factors can be especially important on battlefields with relatively homogeneous geology (one rock or sediment type with little tilt to the strata). The most important topographic features on the Seven Days battlefields were the location of rivers and streams, and the resulting fluvially dissected terrain and poorly drained swamps. Without stream erosion, the Coastal Plain battlefields of Seven Pines, Gaines' Mill, and Malvern Hill would have little change in relief across the landscape.

Common soldiers and commanding officers alike would have had no true understanding of the complex interactions that created the ground they were battling over in the 1860s. The first true geomorphological model was devised twenty years after the war had ended by an American geographer, William Morris Davis. Davis based his model of landscape development and evolution on the concept of uniformitarianism—the idea that physical processes are consistent through time.[5] If small ripple marks, for example, form today on the bottoms of streams and rivers, then an ancient rock that has ripple marks on its surface can be presumed to have been created in a similar environment. If lava is observed to cool into a black igneous rock called basalt, then basalt from an unknown age or source can also be accepted to have once been molten lava, cooling and crystallizing at the earth's surface.

The key to Davis's model, in addition to uniformitarianism, was the consideration of the interplay between long-term uplift of a region and downcutting created by erosion. The primary erosional agent in this scenario is water, and runoff from rivers and streams slowly dissects an uplifted region, perpetually reducing and rounding ridges. Given enough time, according to the model, a battlefield like Kennesaw Mountain will be transformed into a landscape as flat as that of Bentonville (see figure 2.2). Modern geomorphological studies can be broken down by erosional processes, including mass wasting, fluvial (rivers and streams), glacial, aeolian (wind), and mechanical and chemical weathering.[6] Most Civil War battlegrounds can be categorized by the most important erosional processes that contribute to the local landscape. Vicksburg, for example, is underlain by loess, which is windblown silt (aeolian), and the landscape has been drastically modified by the Mississippi River and its tributaries (fluvial). At Antietam, the landscape has been altered by streams (fluvial) and chemical weathering of the carbonate rocks. The same factors influence the landscape at Chickamauga, Nashville, Franklin, and Stones River. Gettysburg has stream erosion as well, but mechanical weathering (freeze-thaw action) is more important than chemical weathering in creating outcrops like Devil's Den or Little Round Top.

Rock Weathering and Terrain

At the surface of the earth, between the hard rocks below and the atmosphere above, is a thin layer of chemically altered rock fragments and

FIG. 5.1. Unusual rectangular exfoliation cracks on a 2-meter long diabase boulder on the summit of Little Round Top at Gettysburg. Each rectangle is approximately the size of a shoebox. Library of Congress.

organic matter. This layer is created by a process called weathering, and the composition of the resulting regolith is dependent on the interaction between the parent material (bedrock) and the climate.

Three primary types of weathering act in conjunction to alter the surface of the planet. Physical (or mechanical) weathering, chemical weathering, and biological weathering act together to modify the bedrock into regolith and soil, and each type of weathering can enhance the effectiveness of the other agents. For example, mechanical breakdown of a rock, in which it is pulverized into smaller fragments, increases the overall surface area of the components, and this process in turn increases the area over which chemical weathering can occur.

Physical weathering takes place when rocks are broken down into smaller pieces by abrasion, expansion, or crushing. During this process of disaggregation, the pieces of rock change size but not composition. When rocks expand they tend to fracture, so deeply buried rocks will break when the overburden above them, and thus the confining pressure, is lessened. Rocks can also be heated by the sun, a process called insolation, and fracture. On many Civil War battlegrounds from the Mid-Atlantic, cyclic expansion from freezing and thawing causes substantial mechanical weathering (see figure 5.1). Exfoliation, which occurs when rocks fracture in an onionlike pattern, produced the rounded boulders of Devil's Den and the Round Tops at Gettysburg.[7] This weathering is enhanced by hydrofracturing, in which water penetrates into the diabase outcrops and freezes and expands, further disaggregating the rocks.

Chemical weathering occurs when chemical bonds within the rock are broken, usually through exposure to acid in solution. Rainwater is the primary chemical weathering agent at the surface, or when it infiltrates into the regolith. Stable minerals in the parent rock are transformed by this weathering process into secondary minerals like clays and hydrous oxides. For silicate minerals such as pyroxene, feldspar, and quartz, the vulnerability to chemical weathering is related to the temperature in which the minerals originally crystallized. The first minerals to crystalize deep in the earth at the highest temperature are more susceptible to chemical weathering after they solidify. Rocks that crystallized at a lower temperature are less vulnerable to chemical weathering. Thus, olivine and pyroxene, two high-temperature ferromagnesian silicates, are less chemically stable than lighter-colored non-ferromagnesian silicates like feldspar or quartz. In simpler terms: dark igneous rocks tend to weather more quickly than those composed of light-colored minerals (the pink granite of Richmond is slightly more durable than the black igneous rocks of Atlanta).

Two Civil War battlefields stand apart as being underlain by particularly hard rock: South Mountain in Maryland and Kennesaw Mountain in Georgia. Despite the inherent durability of the rocks at these sites, the quartzites at South Mountain are slightly less vulnerable to chemical weathering than the pyroxene-rich migmatites that compose Kennesaw Mountain.[8] Nevertheless, both battle sites are famous because of their prominence above the surrounding terrain, and this increased elevation is more a result of structural geology than mineralogy.

Rainwater is slightly acidic (pH ~5.7) and dissolution of carbon dioxide in soil will leave soil water slightly acidic as well. The solubility of CO_2 is inversely proportional to the temperature, so slightly colder climates will generally have somewhat more acidic pore water. When this acidic water comes into contact with calcium carbonate ($CaCO_3$), the calcite compound undergoes dissolution. Landscapes can be transformed with significant amounts of dissolution, producing caves, sinkholes, and other karst features like karrens.[9]

Calcium carbonate is the primary component of carbonate rocks like limestone and dolostone. This category of rocks underlies many Civil War battlefields, including Stones River, Chickamauga, Nashville, Franklin, Antietam, Monocacy, Cedar Creek, and Winchester (Opequon; see figure 5.2).

Biological weathering can increase the rate of physical and chemical weathering. Organisms can change the pH of the environment (increase soil acidity) or break up rocks with their roots or burrowing. Bio-

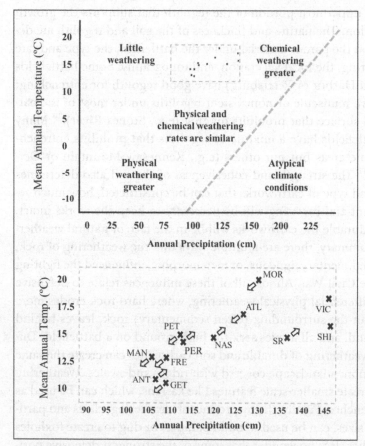

FIG. 5.2. The ratio of chemical and physical weathering is related to the climate (temperature and precipitation) of a battleground. Battlefields from North America cluster between the two primary weathering agents (dashed-line inset) and the arrows indicate a natural shift in the primary weathering type because of the local geology (i.e., Stones River [SR] is shifted towards "chemical weathering" because it is underlain by carbonates). MAN = Manassas; ANT = Antietam; FRE = Fredericksburg; GET = Gettysburg; PET = Petersburg; PER = Perryville; NAS = Nashville; ATL = Atlanta; MOR = Morris Island; VIC = Vicksburg; SHI = Shiloh.

turbation, the mixing of soil and sediments by animals, can create new pathways for water or disturb soil horizons.

Perhaps the most important result of weathering of any form on the fighting during the Civil War was the alteration of the regolith. The quality and quantity of regolith above the bedrock will be a crucial factor in determining what varieties of earthworks can be constructed. The soil is the uppermost portion of the regolith that supports the growth of vegetation. The nature and thickness of the soil and regolith are dependent on the parent material under the battlefield, the type and rate of weathering, the local vegetation, and topography. Some battlefields (e.g., Cold Harbor or Petersburg) have good regolith for entrenching; others have miniscule or nonexistent regoliths under most of the battleground surface that prohibits digging (e.g., Stones River).[10] Many other battlefields have a mixture of regoliths that prohibits entrenching in some areas but not others (e.g., Kennesaw Mountain or Gettysburg).[11] The strength and cohesiveness of the soil also determines the size and type of earthworks that can be constructed, how much reinforcement and revetting will be necessary to keep the works intact, and how durable the earthworks will be in the face of natural weathering. In summary, there are multiple ways that the weathering of rock, whether sedimentary, igneous, or metamorphic, influenced the fighting during the Civil War. Almost all of these influences relate to defensive tactics. Differential physical weathering, where hard rock erodes more slowly than the surrounding (often sedimentary) rock, leaves behind high ground, and all armies seek the high ground on a battlefield. Differential weathering of durable and soft carbonates can create the same phenomenon: a landscape covered with ridges and swales. Weathering can also create smaller-scale features like karrens, which can be used as natural trenches. Finally, the by-product of weathering, clasts and particles of all sizes, can be used to build parapets or dug to create foxholes and trenches. It is no wonder that some of the strongest defensive positions from the war, like Lee's fortified position at Fredericksburg, were cut directly into high, soft, sedimentary rocks.

CHAPTER 6

Killer Carbonates

Chemical Weathering and Rolling Terrain

Sedimentary rocks underlie the majority of Civil War battlefields when compared to igneous or metamorphic rocks. These sedimentary battlegrounds can be divided between those underlain by sediments and clastic sedimentary rocks (sand, sandstones, silt, siltstones, clays and shales) and carbonates (primarily limestones and dolostones). The first subdivision encompasses all the battlefields from the Coastal Plain (Cold Harbor, Seven Days, most of Petersburg, Bentonville) and the Triassic rift basins of the Piedmont (Manassas and Gettysburg). The second sedimentary category, including carbonate rocks, represents even more fields of conflict, including Antietam, Stones River, Chickamauga, and Monocacy, as well as those from the Great Valley (Opequon, Cedar Creek, Kernstown, Cross Keys, and Port Republic).

Carbonate rocks are created in many different depositional environments, including reefs and carbonate marine platforms. Most sedimentary rocks like sandstone and shale are composed of detrital particles (sand, silt, clay), but the grains that compose most carbonate rocks are different: they were once part of something that was alive. Skeletal fragments and tests of creatures such as mollusks, corals, echinoids, foraminifers, and coccolithophorids are made of calcite. When they are compressed and cemented together they form limestone ($CaCO_3$) or dolostone ($CaMg(CO_3)_2$). Carbonate rocks may also be composed of ooids or calcite or aragonite fragments of other creatures, and they may be "enriched" with harder materials like chert.[1] These harder minerals

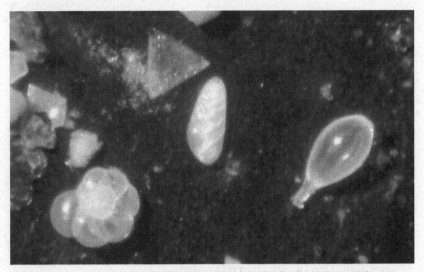

FIG. 6.1. Four microfossils from the sediment from within the Confederate submarine *H.L. Hunley*. The triangular shaped specimen at the top is a diatom, while the three tests below are calcareous foraminifers (*Ammonia* sp., *Bolivina* sp., and *Lagenina* sp.). These microfossils are all approximately 0.25 mm in length.

are usually made of silica and often include siliceous skeletal elements like sponge spicules or diatom frustules (see figure 6.1).[2]

Other limestones form from chemical precipitation of calcite that has been dissolved in ocean or fresh water. Thus, limestone can form massive blankets of thick sedimentary rocks on the continental shelf or thinner layered and banded rock in freshwater environments like streams or lakes. This terrestrial aquatic environment will produce the rock travertine.

The most important quality of limestone, from a geomorphological perspective, is the rock's susceptibility to chemical weathering. Limestone is at least partially soluble, so even slightly acidic water can cause chemical alteration. The result of extensive weathering is karst topography, where caves and sinkholes mark the landscape. The Great Valley of Virginia has multiple carbonate battle sites but also many famous caves, including Luray, Shenandoah, and Grand Caverns.

The most interesting, or at least picturesque, carbonate landscapes from the Civil War is found in the Dry Tortugas (see figure 6.2). These islands, once part of a great Pleistocene reef system, contain the massive brick fortification named Fort Jefferson.[3] Today the carbonate fragments of the eroded reef are the primary component of the beach sands

FIG. 6.2. Fort Jefferson is located on Garden Key, 70 miles west of Key West in the Gulf of Mexico. The carbonate sand surrounding the fort is composed of very coarse fragments of shells and corals. The breakwater, right, served a dual function: It protected the fort from waves and created a moat. Damage from Hurricane Irma (2017) is clearly evident.

in the region. The Confederates never attempted to capture Fort Jefferson and its deepwater port during the Civil War, but the fortification does have a Civil War legacy: several members of the Lincoln assassination conspiracy, including Samuel Mudd, Samuel Arnold, Edmund Spangler, and Michael O'Laughlen, were imprisoned on the island in the late 1860s.

Carbonate Rocks and Combat

Four battlefields, two from each theater of combat, demonstrate the influence of carbonate rocks on the terrain and tactics of the Civil War. In the West, Stones River and Chickamauga were bloody, indecisive battles where limestone aided the Union army's survival to fight again. In the East, the combination of softer limestones and shales and the more erosion-resistant dolostones benefited (qualified) Federal victories at Antietam and Cedar Creek. The influence of carbonate rocks on the combat and tactics of these battles will be explored over the next several pages, with case studies comparing the fighting across these limestone landscapes.

ANTIETAM AND STONES RIVER

Stones River was an indecisive battle during the fight for Tennessee, one of the first large clashes in the Western Theater. Antietam was an indecisive battle in Maryland, a key battle in the Eastern Theater during the second year of the war. Despite these geographic differences, the geology of the Stones River and Antietam battlefields is remarkably similar, and the influence of the carbonate rocks on the fighting during the 1862 battles is comparable in many ways. Both battlefields are primarily underlain by relatively pure limestones that produce gently sloping, undulating terrain. At Stones River this carbonate formation is known as the Ridley Limestone; it covers more than 80 percent of the battlefield and is known for producing karstic terrain.[4] At Antietam the majority of the battlefield, and the flattest terrain, is underlain by the Conococheague Limestone, a rock very similar in composition and texture to Stones River's Ridley Formation.[5]

Both Stones River and Antietam have a smaller portion of the battleground underlain by more heterogeneous carbonate rocks. Along the Stones River, the Pierce and Murfreesboro Limestones underlie more rolling terrain. These formations have a variable clastic content (mostly clay) and are interbedded with shales.[6] The result of this variability in composition is a rolling terrain, with harder portions of the strata producing ridges on the landscape and fords across the river. At Antietam, the eastern and southern portion of the battlefield lies above a similar heterogeneous carbonate formation, the Elbrook. These Cambrian strata are composed of interbedded and repeating layers of limestones, dolostones, and shales. The weaker shales tend to produce valleys or swales while the dolostone produces ridges.

To briefly recap the similarities in the geology surrounding Murfreesboro and Sharpsburg: the western and northern portions of the battlefields are underlain by purer limestones that produce flat or gently undulating terrain, and the smaller eastern portion of the fields are underlain by a mixture of carbonate and clastic rocks that produce rolling terrain. Undulating terrain provides little cover for attacking soldiers, especially where forests are scarce, and the flatter landscape increases the effectiveness of artillery: sight lines are better and solid shot

FIG. 6.3. The northern sector of the Antietam Battlefield, where the fighting was heaviest in the morning, is underline by limestone that creates a flat or undulating terrain (left). The more durable Elbrook Formation (right) creates a rolling terrain. Burnside Bridge is found below this tough, outcropping dolostone boulder.

and canister have a better probability of ricocheting or grazing (see figure 6.3).[7] Rolling terrain, in contrast, provides ridges that are preferential for defensive lines, but the swales will provide cover for attacking columns of infantry. High ground is terrific for defense—especially with reverse-slope firing positions—unless there is additional high ground immediately to the front of the position. Parallel ridges provide concealment for approaching infantry, reducing the range of the engagement to the area between hillcrests.

Both battles progressed in an identical manner: the opening phase of the battle was fought across generally flat terrain above purer limestones (the Conococheague and Ridley Formations) while the later phases of the battle shifted to rolling terrain of the Murfreesboro and Elbrook units. On a smaller scale, however, the geology of the battlefield indirectly influenced the fighting in different, subtler ways.

The soil at Antietam is much better developed and thicker than that at Stones River. While the Conococheague crops out at several locations north of Sharpsburg, the Elbrook remains largely hidden under a thick blanket of rich farm soil. At Stones River rock outcrops are much more common. The thin regolith resulted in two characteristics of the battleground that were important for battle: thin soils prevent tillage, and thus areas that are stony tend also to remain wooded. Forest cover limits sight lines for artillery and offers some degree of protection for infantry on the defensive. Wooded terrain also reduces the range of engagement for infantry and artillery. Thin regolith additionally prevents entrenching. When Confederate general Braxton Bragg formed his ini-

tial defensive line north of Murfreesboro when the battle was imminent, only approximately half his line could be dug in because of the
shallow limestone.

Perhaps the most important karstic feature on either battlefield,
in terms of influencing combat, were the karrens (or "cutters") found
at the center of the Federal line at Stones River.[8] These natural, one-
meter-deep trenches offered protection for the infantry of Union generals Sheridan and Negley as they withstood multiple rebel assaults. The
Union troops held this position, anchoring the Federal line in bedrock,
until ammunition ran low. The time occupied by these repeated and
costly Confederate assaults, and the delay in bringing forward cannon
through the limestone obstacles, provided Maj. Gen. William Rosecrans
time to reform his new and final defensive line to the north, protecting
his only viable supply line and potential path of retreat, the Nashville
Pike. Without the defensive benefits of the karrens and the many limestone obstacles to the Confederate advance, Bragg might have been
able to follow the retreating Federal line and crush Rosecrans's fallback
position, essentially destroying the Army of the Cumberland.

At Antietam, the Conococheague Limestone produced a relatively
flat, open terrain that was covered with cornfields and pasture land.
Multiple Federal assaults and counterattacks by Stonewall Jackson's
Second Corps led to especially high casualty numbers. At Stones River
the ground is similarly gently sloping and undulating, but the thin soil
limited farming, and the surviving forests provided cover and concealment for troops on the move. At Antietam, much less of the landscape
is wooded. As a result, during the morning assaults of both battles, the
range of artillery and small arms firing was greater at Antietam than at
Stones River.

In summary, the limestone around Sharpsburg strongly favored the
defense during the first phase of the Battle of Antietam, while a similar limestone around Murfreesboro favored both offensive and defensive tactics during the Battle of Stones River: karrens made a key defensive stand possible and limited the mobility of Confederate artillery
once the position was abandoned, but the thin regolith prohibited the
construction of trenches and left forests intact to both increase concealment and diminish the effectiveness of artillery.

The final phases of the Battles of Stones River and Antietam were also
fought across similar carbonate landscapes with similar tactical implications. After Burnside eventually spanned the bridge that would later

bear his name, his assaults toward the Confederate right wing crossed the challenging, rolling terrain of the Elbrook Formation. During the final day of the Battle of Stones River the Confederates launched an assault against the isolated left wing of the Federal line. This assault paralleled the river and traversed the rolling Pierce and Murfreesboro Limestones. During both assaults the effectiveness of long-range artillery and small arms fire was slightly diminished by the nature of the terrain; approaching infantry would remain hidden in the swales before rushing quickly across the higher ground before seeking shelter again in the next proximal valley to the enemy.[9] As a result, defensive fire was not truly effective until the last hillcrest was approached, at which point the firefight would take place at a fairly close range.

An account after Antietam by Private Alexander Hunter of the Seventeenth Virginia demonstrates the challenge of holding a defensive line on a rolling landscape:

> Each man sighted his rifle about two feet above the crest, and then, with his finger on the trigger, waited until the advancing form came between the bead and the clear sky behind. The first object we saw was the gilt eagle which surmounted the staff, and after that the flutter of the flag itself; slowly it mounted, until the Stars and Stripes were flying all unfurled before us. Then a line of hats came into sight, and still rising the faces beneath them emerged and a range of curious eyes were bent upon us; and then such a hurrah as only Yankee troops could give broke upon our ears and they were rapidly climbing the hill and surging towards us. "Keep cool, men, don't fire yet!" Colonel Corse shouted, and such was the perfect discipline that not a gun replied; but when the Yankee band flashed above the hill-top the forty-six muskets exploded at once and sent a leaden shower full into the breasts of the attacking force, who were not over sixty yards distant.[10]

It is a commonly held trope that the American Civil War was especially bloody because the commanding officers failed to comprehend the influence of the deadly new rifle musket in combat. In theory, the new weapon could hit a man-sized target consistently out to a range of 400 yards, while the earlier Napoleonic-era smoothbore muskets were only reliable to hit a similar target at 100 yards. In reality, most infantry vs. infantry firefights took place at much shorter ranges.[11]

Perhaps the largest reason rifle muskets were not fired en masse at the enemy at ranges between 150 and 400 yards was geology. In the two battles previously discussed, the terrain would really only provide one scenario when long-range rifle fire might be effective: during the morning phase at Antietam fought across the Conococheague Limestone. During the initial phase of the Battle of Stones River the forest cover resulting from the thin regoliths would limit visibility and long-range fire. During the later stages of both battles, the fighting occurred across heterogeneous carbonates that created rolling terrain, and rolling terrain requires the initial or primary combat to take place at shorter ranges. Rifles were not fired at their true potential range because of rocks, but also because of the limit on visibility provided by gun smoke and the desire to conserve ammunition when preparing to receive a massed assault. Nevertheless, these final two factors prove inconsequential when the geology of a region renders an approaching enemy invisible.

CHICKAMAUGA AND CEDAR CREEK

Harder carbonate rocks at the Battles of Chickamauga and Cedar Creek produced higher ground favorable for defense and saved two different Federal armies from total defeat. The carbonate rocks along the Chickamauga Creek in Georgia are similar in age, composition, and original depositional environment to those found along the Stones River, 100 miles to the northwest. Maj. Gen. Rosecrans selected a position on slightly higher ground than his adversary, on rocks that were more erosion resistant because of the higher chert content in the limestones.[12]

The first day of the Battle of Chickamauga saw no clear winner, but the armies collectively amassed more than fifteen thousand casualties. On the second day of fighting, the Federal line collapsed after miscommunication between the commanding officers left a critical gap in their defensive line. The Union line was eventually reorganized by Gen. George "The Rock of Chickamauga" Thomas along the heights of Snodgrass Hill and Horseshoe Ridge.[13] This position is underlain by the hardest, silica-enriched carbonates on the battleground, and this geologically enhanced position would permit the Federal infantry and artillery to survive twenty-five rebel assaults before they quietly left the battlefield after sunset.[14]

Harder carbonate rocks similarly provided higher terrain for the Federal forces at the Battle of Cedar Creek in Virginia, slowing Confederate momentum to the point that the Union army could launch a counterattack, reversing the probable outcome of the battle. The Battle of Cedar Creek was the final battle of Lieut. Gen. Jubal Early's Valley

FIG. 6.4. At the Battle of Cedar Creek the initial Confederate assault swept over the pastureland and cornfields south of Middletown, Virginia. Later in the day, the southern infantry would flee south across this same carbonate landscape.

Campaign of 1864. He launched a surprise early morning attack across Cedar Creek, using the rolling terrain created by shales and sandstones of the Martinsburg and Oranda Formations for concealment. This assault was completely successful, driving several Union infantry divisions out of their camps and back across the flat fields above the Edinburg, Lincolnshire, and New Market Limestones (see figure 6.4).

The Confederate onslaught was only slowed when it reached the outskirts of the town of Middleton; here Union brigadier general George Getty formed his defensive line across a cemetery situated on high ground. This line paralleled the contact of the softer limestone to the south and west and the harder, erosion-resistant dolostone of the Beekmantown Group. The Federal division positioned on the dolostone withstood assaults across the lower limestones by four Confederate divisions for more than an hour. This stubborn defense slowed the southern momentum, confusing Early, who believed he was attacking an entire Union corps instead of a single division, and this lull in the fighting allowed the Federals to form a new defensive line to the north of town (see figure 6.5).

FIG. 6.5. Federal Private Robert Knox Sneden's sketch "Position of Union Army at Cedar Creek; Union position and plan of battle." Federals occupy the higher dolostone (right), after being pushed back across limestone during the first half of the battle. Enhanced detail of the sketch from the Library of Congress.

By late morning Early and the smaller Confederate army had nearly completed a tremendous tactical victory: they had captured almost 1,500 Union soldiers and more than twenty badly needed cannon. The two armies were now a half mile apart, with the Confederates positioned on limestone and the Union on harder, higher dolostones. At 4:00 in the afternoon Maj. Gen. Philip Sheridan, who had just arrived on the field, launched a counterattack with Wright's Sixth Corps and Emory's Nineteenth Corps. Each flank of the two-corps attack was protected by a cavalry division. The Confederate line held strong for an hour before the left flank began to weaken and Brig. Gen. George Custer's cavalry was able to swing south into the rear of the rebel infantry. This maneuver caused a general panic in the southern line, as the potential escape route south across Cedar Creek was now imperiled. As the routed Confederate army began the rushed retreat, they left behind many of the captured guns and wagons taken during the morning's success. At the end of the day, Sheridan and the Union army had completely reversed the tide of the battle and held the carbonate battlefield, but at a cost of more than 5,500 men. The Confederates lost around 2,500 fewer men, but their final invasion toward Northern territory was embarrassingly ended, as was Early's career as a commanding officer.

In summary, then, differential weathering of carbonate rocks produced landscapes that were exploited by skilled Federal commanding officers in the field, supporting both offensive and defensive tactics. High ground from harder dolomite and chert-enriched rocks abetted several important defensive stands, while weaker rock produced swales that provided concealment for infantry on the move. Karst features were also used to provide a force multiplier on the defensive, such as with the sinkholes and karrens at Stones River. A subtler, indirect influence of limestone on battlefield terrain involves land-use decisions. On several battlefields, including Antietam and Stones River, the thin, gravelly soil above the carbonate basement rocks did more than prohibit entrenching: the lack of soil made the land unsuitable for farming and thus forests survived into the 1860s. These woods provided some degree of cover for offensive maneuvers, although dense forests inhibited coordination across the battlefield. The trees also provided cover for infantry in defensive positions and limited the mobility and effective range of artillery. Finally, thick forests dictated that infantry engagements occur at a shorter range when compared to other terrain, further diminishing the value of the rifle musket over the smoothbore. In a small way, this influence on the range of engagements may have benefited the Southern soldiers, who were more often armed with older, less accurate smoothbore weapons, especially during the first years of the war.[15]

CHAPTER 7

Battling in the Basins

Heterogeneous Landscapes

The most interesting battlefield landscapes, from a tactical or geological perspective, are those underlain by rocks of different origins or compositions. Most larger, well-known terrain features from the Civil War are found where battlefields offer a variety of rock types: Little Round Top (diabase) is surrounded by sedimentary strata of the Gettysburg Formation; Kennesaw Mountain (hard migmatites) is surrounded by slightly less durable igneous and metamorphic rocks; Lookout Mountain (hard sandstone) is underlain by weaker carbonates; Stony Ridge at Manassas (diabase) is surrounded by softer sedimentary rocks of the Bull Run Formation. This relationship between rock heterogeneity and tactical and historical importance is not coincidental: hard rocks surrounded by softer rocks results in high ground, and high ground is preferential as a defensive position—and this terrain invariably ends up as the key to controlling the battlefield.[1]

Variable Erosion and Defensive Stands:
Second Manassas and Gettysburg

Although the Gettysburg and Manassas battlefields are 70 miles apart, they are both located in Triassic rift basins within the Piedmont. As such, they have nearly identical rock formations. At Manassas, the majority of the battlefield is underlain by Triassic shales and sandstones of the Bull Run Formation, and specifically a portion of the unit called the

Groveton Member. Sedimentary structures within this formation include mud cracks and ripple marks, suggesting a lacustrine (lake) shoreline as the original depositional environment for these sediments.[2] These layers of sedimentary rock are part of the massive wedge of sediment that filled the Culpeper Basin during the first period of the Mesozoic era.

During the next period of geologic time, the Jurassic, massive rifting in the region associated with the breakup of the supercontinent Pangea allowed molten material to rise from below and crystallize as igneous intrusions of a dark (mafic) rock called diabase.[3] These intrusions altered the preexisting sedimentary layers with their heat, baking the sandy and silty strata to produce contact metamorphism and a resulting altered rock called hornfels. After cooling, then, each igneous intrusion or dike is surrounded by a 20-foot-wide ring of metamorphic rock. In some places the diabase also shows evidence of variable cooling rates, with smaller crystals forming where the dike cooled quickly along the edges and larger crystals forming at the core of the dike, where slower cooling resulted in the formation of larger mineral crystals.[4]

The geology of the Gettysburg battlefield is nearly identical. Instead of the Bull Run Formation, the red and brown sedimentary Gettysburg Formation underlies most of the northern portion of the battleground. This thick section of the Gettysburg Basin also dates to the Triassic, and sedimentary structures and trace fossils (dinosaur footprints) suggest a lakeside depositional environment. Around 200 million years ago molten material intruded into these sediments, producing diabase dikes and sills, and, just as at Manassas, a zone of contact metamorphism.[5]

While the geology and formation of the Manassas and Gettysburg battlefields is similar, there are also subtle differences in the landscape. Diabase outcrops are found on both battlegrounds, but the rocks are much easier to study at Gettysburg because of the tactically important terrain features like the Round Tops, Devil's Den, the Slaughter Pen, and Culp's Hill. Larger spheroidally weathered boulders are also more common on the Pennsylvania battlefield. Contact metamorphism is more prevalent at the surface and easier to observe at Gettysburg as well, especially along the (in)famous railroad cut north of town that proved so important during the first day of fighting.

Even though the geology of the two battlefields is very similar, the role of the rocks in the outcome of the battles is more variable. The diabase at Manassas needed to be modified by human hands before it proved tactically important to the Confederates. Jackson established

FIG. 7.1. Part of the reason Cemetery Ridge was not entrenched is this sill, which crops out on the Bryan Farm. Seminary Ridge is visible across the open fields over which Pickett conducted his doomed charge.

his defensive line on diabase along an unfinished railroad grade—a strong defensive position that Union general John Pope launched disjointed attack after attack against. Diabase helped the defending Confederates, even to the point of serving as projectiles.

At Gettysburg, a similar defensive orientation occurred. Meade established his defensive "fishhook" line on diabase, and Lee launched assaults on the second and third day of battle from one igneous body, Seminary Ridge, across a sedimentary landscape toward a larger igneous body, the Gettysburg Sill (Little Round Top, Cemetery Ridge, Culps Hill) (see figure 7.1).

Comparing Geology and Tactical Choices

Examples of the influence of sedimentary geology on battlefield tactics are most easily observed when geographically proximal battlefields are compared. When Lee ordered Pickett, Pettigrew, and Trimble to charge toward Cemetery Ridge at Gettysburg, his men needed to cross a broad, gently undulating landscape underlain by soft sandstones, siltstones, and shales. While this assault proved disastrous, a similar attack might have been successful on the Antietam battlefield, located only 40 miles to the south. At Antietam, the attack would have been conducted on a completely different type of sedimentary geology, as the landscape around much of Sharpsburg is underlain by carbonate rocks, which weather in a much more inconsistent manner compared with those of the Gettysburg Basin. As a result, the fighting around the Sunken Road or Lower Bridge at Antietam took place on rolling terrain, and

FIG. 7.2. Union position on Cemetery Ridge looking towards Seminary Ridge. Lee's men emerged from the distant forest to begin the infamous Pickett's Charge. During the attack they marched from a diabase dike across farm fields underlain by sandstones towards this rock wall, positioned on a diabase sill.

FIG. 7.3. View along the Sunken Road (later "Bloody Lane") at Antietam. Had Pickett's Charge taken place on similar terrain, the variability in weathering rates between limestones (carbonates) and the resulting rolling landscape would have hidden the approaching Confederates until the soldiers emerged on the harder dolostone hillcrest less than 100 yards distant.

these ridges and swales were the result of differential weathering of limestones and dolostones. Any massive assault—like Longstreet's at Gettysburg—across this terrain would have left the approaching rebel lines invisible to the defending Federals. The rolling topography would have hidden the approaching columns and greatly diminished the effectiveness of the Union artillery. As figures 7.2 and 7.3 document, Lee's men would have been shielded by the rolling carbonate terrain at Antietam, and the Union defenders would have had a limited amount of time to fire on the advancing wave of enemy soldiers as they crested the final hill less than 100 yards away. Instead, at Gettysburg the Confederate infantry needed to cross a flat field underlain by sandstone, siltstone, and shale, all of which eroded in a manner to produce a flat, open, and deadly landscape.

Geomorphology, Weathering, and Terrain

This section of the book discussed the vast array of ways that geology and rock weathering, both chemical and mechanical, could influence the lay of the land; these geomorphological phenomena ranged from differential weathering between rock types to fluvially incised stream valleys. Skilled commanders used the resulting landscape features in a variety of ways when conducting both offensive and defensive maneuvers. On nearly all battlefields, the best terrain for defensive positions was above the hardest rock, and often on hard rock that had been uplifted and tilted. And, high ground on hard rock isn't necessarily the strongest position for a defensive line because the thin regolith above the resistant rock prohibited digging: Little Round Top, for example, was fortified by piling large sediments, not digging into them. Lee's line at Fredericksburg was an exception to this rule about hard rocks and high ground; his line was positioned on high ground underlain by softer rocks that could also be entrenched, a twofold bonus for infantry and artillery. Moreover, this defensive position, with all of its geological enhancements, was probably the strongest occupied during the war. Lee summed up the value of the earthworks well on December 14, 1862: "My army is as much stronger for these new entrenchments as if I had received reinforcements of 20,000 men."[6]

CHAPTER 8

Sedimentary Geology and Logistics

Military Logistics

The art of effectively moving and supplying an army in the field falls
under the realm of military logistics.[1] For an army to produce a successful field campaign, it must be properly supplied when in camp and on
the move. After all, a rifle musket without adequate ammunition is essentially an elaborate club, and a hungry soldier is not a motivated soldier. The influence of sediments on the movements of infantry, artillery,
and supplies is obvious when considering the numerous times maneuvers became, quite literally, bogged down along muddy roads. Nevertheless, geology produced a profound effect on the logistics of the war
on a much broader scale: the location of militarily important cities, supply bases, foundries, and arsenals—especially in the South—are all the
result of a shift in rock type across the landscape.

The Military Heritage of the Fall Line

Sedimentary rocks are usually less durable and easier to erode by water than metamorphic or igneous rocks. Unconsolidated sediments are
always the least resistant of these geological materials and are the most
susceptible to weathering and erosion. When a river flows across a contact between harder igneous or metamorphic rocks and partially lithified or unconsolidated sediments, there is nearly always an abrupt
drop in elevation; the sediments are eroded more quickly and rapids
or waterfalls are created at the edge of the harder rock. This is exactly

the geological circumstance along the extent of the contact between the harder crystalline rocks of the Piedmont and the weakly lithified sedimentary rocks and sediments of the Coastal Plain.

The Atlantic Coastal Plain began to form around 100 million years ago as sediments shed from the actively eroding Appalachian Mountains were deposited onto the margin of the newly expanding Atlantic Ocean. Sea level rise and fall over the millennia produced thousands of feet of sedimentary strata adjacent to and above the much older igneous and metamorphic rocks of the eastern Piedmont. Modern rivers flow from the uplands to the north and west across the Piedmont before crossing the Coastal Plain to the Atlantic and Gulf coasts. When these rivers traverse the Piedmont/Coastal Plain contact, erosion is greater downriver, producing a series of waterfalls and a zone of rapids. Taken collectively, the waterfalls along the river form what is termed the Fall Line, or perhaps more properly Fall Zone—the line across the landscape that marks the shift from hard rock to softer rock, and the start of the broad, flat, Coastal Plain.[2]

Before the construction of bypass canals, river travel was seriously hindered by the rapids of the Fall Line. Moving upriver across the Coastal Plain was relatively effortless, but continuing upstream across the Piedmont was impossible without a means of circumventing the waterfalls. Thus, a change in transportation mode was required to negotiate onto the hard-rock region, and the location where disembarkation was required quickly began to flourish as communities began to grow. Soon roads, and later railroads, emerged from these Fall Line villages, and as the importance of the transportation hub grew, so followed population. As a result, many important cities lie along the Fall Line, including the capitals of both belligerent factions during the Civil War (see figure 8.1).

The increased gradient of the river at the Fall Line, and the resulting higher-velocity water flow and energy, also led to the introduction of industry to this portion of the river channel. Multiple Confederate foundries, including the Tredegar Iron Works on the James River at Richmond, owe their site selection to the Fall Line and weak sedimentary geology of the Coastal Plain (see table 8.1).

The city of Columbus, Georgia, straddles the Fall Line along the Chattahoochee River. The only city in the South to offer more war matériel to the Confederacy was Richmond. Factories lined the rap-

FIG. 8.1. Granodiorite exposed along the Great Falls of the Potomac. Over this stretch of the Potomac the river falls almost 80 feet in less than a mile.

TABLE 8.1. Military significance of important Fall Line cities of the Confederacy

Note that this list omits smaller Fall Line towns that also had a military significance, including sites of battles and railroad junctions (e.g., Hanover, Va., Weldon, N.C., Raleigh, N.C.).

Confederate Fall Line	River	Military significance
Fredericksburg, Virginia	Rappahannock	Fredericksburg CSA Shipyard
Richmond, Virginia	James	Tredegar Ironworks (artillery), S. G. Robinson Arms Manufactory (carbines), Richmond Armory (carbines), Union Manufacturing Co. (rifles), Virginia Manufactory (rifles)
Rocky Mount, North Carolina	Tar	Rocky Mount Mills (yarn and cloth)
Fayetteville, North Carolina	Cape Fear	Fayetteville CSA Naval Ordnance Depot, Fayetteville Arsenal (rifles)
Columbia, South Carolina	Congaree	William Glaze and Co. ("Palmetto Armory"; rifles), Congaree Foundry (artillery), Palmetto Iron Works (rifles)
Augusta, Georgia	Savannah	Morse Co. (carbines), Beech and Rigdon (rifles)
Milledgeville, Georgia	Oconee	George State Armory (rifles, ammunition)
Macon, Georgia	Ocmulgee	Hodgekins Co. (rifles), Dickson, Nelson, and Co. (rifles)
Columbus, Georgia	Chattahoochee	J. P. Murray Co. (rifles), Columbus Co. (rifles), Confederate Naval Yard Ironworks
Tallassee, Alabama	Tallapoosa	Tallassee Co. (carbines and rifles)

FIG. 8.2. Columbus straddles the Fall Line along the Chattahoochee River on the border of Georgia and Alabama. At low water most of the rapids are confined to the deepest part of the channel closer to the Alabama bank.

ids along the riverbanks, producing flour, uniforms, and all types of armaments for the rebel cause. Downriver, on the deeper, quieter water of the Coastal Plain, sat the Confederate Naval Yard Iron Works. This manufacturing plant provided machinery and cannon for several important Confederate vessels until it was burned after Wilson's Raiders captured the city on April 16, 1865, during what was arguably the final battle of the Civil War. Today, a short stroll south through Columbus's riverfront park takes visitors past the rapids of the Chattahoochee River and what remains of dams built for companies like Eagle and Phenex (makers of Confederate uniforms) and, downstream, the Confederate Iron Works (makers of small arms and cannon). Walking another quarter mile downstream completes the transition from Piedmont to Coastal Plain, and here along the quieter water is the former site of the Naval Iron Works. At Columbus, the Fall Zone of the river provided the hydropower for manufacturing and the deep water for moderate-draft ships, all within walking distance of the city center and rail depot (see figure 8.2).

The shoreline of the James River in Richmond, Virginia, follows an identical plan along the Fall Zone: the Tredegar Iron Works sits on the rapids of the Piedmont beside the Confederate States Laboratory, an explosives manufacturer, while the CSA naval shipyard at Rocketts Landing resides downriver across the Fall Line on the sedimentary Coastal Plain. The naval yard supplied and serviced the vessels of the James River Squadron, which operated exclusively downriver from the Fall Line. The shipyard constructed casemate ironclads, including the CSS *Fredericksburg* and *Richmond*. Not coincidentally, Rocketts Landing and the naval works at Columbus, Georgia, both ceased operation within two weeks of each other during the last gasps of the rebellion.

The Fall Line cities of Richmond and Columbus were the two principal purveyors of war supplies to the Confederate armies, and the challenge of getting these provisions and arms to the men in rank continued to grow during the war. This was the problem presented to the logisticians of the South: how to move vast quantities of war goods across the Piedmont and Coastal Plain to the armies in the field.

Sedimentary Geology, Transportation, and Logistics

Manufacturing or acquiring military supplies was only the beginning of the task of furnishing the requirements of an army of tens of thousands of men in the field. These supplies, many of which were perishable, had to be transported by rail, ship, and/or wagon to the camps as quickly and securely as possible. In the Confederacy, this transportation was even more difficult because of the poor(er), uncoordinated network of rail lines and the Fall Line: the rivers of the South were essentially cut into two navigable sections by the rapids of the rivers. Military logisticians would need to deal with these challenges to bring food, arms, and ammunition to an army on the move; planners would also need to determine the best way for the army on maneuver to carry the material necessary to fight and survive while marching, even in hostile territory. Here the Union had a significant advantage, as their quartermaster general, the civil engineer Montgomery Meigs, was particularly effective in all aspects of military logistics.

Civil War logisticians planned to supply a soldier with three pounds of food every day, plus an occasional resupply of ammunition, shoes, and clothing. Armies and their supply train are reliant on horses and mules, and one horse was needed for every two or three men in a fighting force—and, of course, these pack animals would need food as well.[3] In total, then, an army of a hundred thousand men would need more than a million pounds of food and grain per day when in the field, and local supplies of sustenance would soon be exhausted during periods of static fighting.

Means of transportation between Fall Line foundries or farmers' fields could include coastal waters and rivers, rail lines, or roads.[4] All were vulnerable to interception, although Federal blockade ships provided some degree of protection for Union supply ships plying coastal shipping lanes.

Early in the war Confederate cavalry raids against Union rail lines disrupted entire campaigns. For example, Grant's 1862 march on Vicksburg via Kentucky and Tennessee down the Mississippi Central Railroad Line was halted when Confederate general Earl Van Dorn led a cavalry raid against his supply base at Holly Springs, Mississippi. The successful attack destroyed most of the army's supplies and severed the rail line, requiring Grant to retreat north to Memphis. Nevertheless, rail tracks used by the Union could often be repaired in less than a week, and the effectiveness of rebel raids diminished during the second half of the war.[5]

The size of the wagon train needed by an army was dependent on the agricultural productivity of the area of operation and the distance from their supply base; if operating far from base, a portion of the wagonload would need to be dedicated to supply the teamsters and mules.[6]

The standard wagon used during the Civil War could carry between 1,000 and 4,500 pounds, depending on the quality of the roads and the weather. A general rule of thumb was that twenty-five six-mule wagons were needed to support a regiment of a thousand men.[7] As a result, 50-mile-long wagon trains were not uncommon during the war. McClellan's wagon train on the Peninsula in Virginia and Sherman's train during the Atlanta Campaign each contained more than five hundred wagons.[8]

The swiftness at which these long trains could travel, as well as their cargo capacity, was dependent on the nature of the roads and the amount of recent precipitation (or lack thereof). An army that outpaced its train support would be especially vulnerable because it would be too far afield from its source of supplies and horse-drawn artillery. As a result, the mobility of the army becomes directly linked to the speed of the supply train and, thus, sedimentary geology: the gravel of paved roads or the compacted soil of unimproved dirt roads directly determined how quickly a fighting force could cross the territory.

Geology and Civil War Roads

During the American Civil War, as today, there were two primary types of roads: hard-surface roads and largely unimproved dirt roads. Unlike today, however, the hard-surface roads during the mid-nineteenth century were constructed out of only two primary materials: crushed local limestone or small-diameter tree trunks or wood planking. For rapid (and more comfortable) travel, the rock-covered road was usually superior.

The technique of lining the surface of a road with pulverized limestone was invented by Scottish engineer John McAdam in the early 1800s.[9] His process consisted of placing three distinct layers of crushed rock (aggregate) on the surface of the dirt roads. The first two layers of hand-crushed aggregate were a little over three-quarters of a foot deep and consisted of what today would be classified as #1 aggregate, angular cobbles averaging 2 or 3 inches in diameter. These clasts allowed for good drainage of the road because of the sediment's great porosity and permeability. The final covering layer of crushed limestone had a finer grain size, similar to #2 aggregate, with particles having a diameter of about one inch. This 2- to 3-inch-thick layer was tamped down by hand or with heavy rollers, forcing the angular cobbles and gravel to compress, enhancing the intergrain friction and increasing the load-bearing capacity of the road. Limestone, a rather soft sedimentary rock, was a good choice of construction material because of the need to break all the rocks by hand; use of a quartz sandstone, for example, would have been exceedingly tedious and produced a roadbed with sharp-edged clasts.

The softness of limestone also meant that traffic by wagons would soon help smooth the angularity of the surface particles of a new road. After months of use the rocks would be rounded enough to allow comfortable travel by men on foot. A road composed of igneous rock aggregate or a material like quartzite would leave a surface of angular, sharp particles with edges that would shred shoes or bare feet alike.

Unfractured limestone is not particularly porous or permeable (porosity < 5 percent), but when pulverized into gravel and cobbles, the porosity and permeability increase dramatically (> 30 percent porosity).[10] Thus, finer-grained limestone gravel was placed in drainage ditches along the sides of the *macadamized road* to enhance the dissipation of water from the rapidly draining roadbed.

FIG. 8.3. A Federal Engineers constructing a corduroy road in the vicinity of Richmond, Virginia, in June 1862. Detail of a stereoview image from the Library of Congress.

The most important macadamized road during the Civil War traversed the Great Valley. The limestone for the Valley Turnpike was quarried all along the 93-mile route between Winchester and Staunton. Stonewall Jackson (fittingly) and his foot cavalry exploited the fine condition of this road, and its relative imperviousness to degradation from weather, to rapidly maneuver up and down the Shenandoah Valley during the Valley Campaign of 1862.[11] Two years later the road would provide a path of rapid transport during Early's raid toward Washington and Sheridan's Valley Campaign.

Although not as permanent, long *plank roads* were also constructed prior to and during the war. After roads were properly graded for water drainage, wood planks, often pine, were laid across wooden sills. These roads were especially common in areas like North Carolina where the supply of lumber for the 8-inch-wide planks was abundant.[12]

During wartime, the need for a road across muddy terrain often prohibited the time-consuming crafting of planks for the road surface. To increase the rapidity of road construction, impassable stretches of dirt roads were simply covered with straight(ish) tree branches and lengths of small tree trunks, creating what was referred to as a *corduroy road* (see figure 8.3). These could be thrown down quickly by soldiers with the proper tree-cutting equipment, but they provided a harsh, punishing surface for wagons or cavalry.[13]

All of these road-covering techniques, of course, are engineered to avoid dealing with the largest of problems created by the smallest of sediment: the combination of clay and silt with water to make mud.

Railroads

Civil War armies relied on railroads for both transportation and supply, and many critical battles were fought for control of rail junctions. Chattanooga, Corinth, Atlanta, and Richmond were all critical convergence points for railroads across the South. As logistics became more important during the war, so did the proper organization and implementation of railroads, which provided a large advantage for the Union. The Southern rail lines were poorly coordinated and organized, constructed of different gauges, and underappreciated by the Confederate leadership. To make matters worse for the Southern logisticians, most of their lines ran north-south across the seceded states. The Confederacy's railroads were, in short, constructed to get cotton to the coast, not to move armies or supplies.

At the start of the war, there were around 30,000 miles of railroad track across America, with about three-fourths of it in the North and almost all of it east of the Mississippi. From a geological perspective, this is a surprise, because building a rail line across the landscapes of the South was a much easier endeavor. The Coastal Plain is much broader in the southern United States, and the soft sediments of which it is composed make for less complicated railroad engineering.

Of key consideration in the construction of the line is the *ruling grade*, that is, the maximum slope encountered by a locomotive during transit between two points. A rail line could be completely flat for 99.5 percent of its path between two cities, for example, which would require only a small locomotive; however, if there is one small stretch of track that has a 4 percent grade, a locomotive will need to be strong enough to overcome this portion of the journey. Thus, it was necessary to construct the grade of the rail line as flat as possible for the entirety of the journey to ensure as large a load as possible could be carried by a train. In general, a grade of less than 1 percent is preferred.

The Coastal Plain has two advantages with regard to the consideration of the ruling grade. First, it is generally very flat, requiring fewer locations that needed either tunnels, tall bridges, or excavation. Second, when excavation is required, sand, gravel, and clay are easier to move than rock.

The Piedmont province presents many challenges when building a railroad. Variations in rock type result in a generally rolling terrain. The Valley and Ridge province and the Appalachian and Blue Ridge Mountains are a nightmare for railroad engineering. Hard rock and high relief make creating a path with a low ruling grade exceedingly problematic. Hard rocks make tunneling and excavation difficult. Even going around mountains or through curving mountain passes is complicated by the nature of railroad engineering: when train tracks curve, more power is needed from a locomotive to overcome the friction of the wheels of the railcars.[14] When steam-powered locomotives enter a long tunnel, a similar need for more power (or a lighter load) arises because the poor ventilation of the tunnel will rob the engine of needed oxygen.

There was one advantage that hard rocks offered to railroad construction. Areas with harder rocks, or at higher elevations, tend to have higher gradients for their rivers and streams. A higher gradient means more vertical erosion and downcutting. Bridges will be needed to cross river gorges, but floodplains will be narrow. A bridge might need to be very tall, but usually it won't need to be especially long. When rivers enter an area with a much lower gradient, like the Coastal Plain or the Mississippi River Valley, they begin to meander. Meandering produces a very broad floodplain, which in turn will require a very long railroad bridge.

Two railroads that were important supply lines during the Civil War demonstrate the interaction between geology and location well. The Cumberland Valley Railroad was a critical line of support for Union troops operating in the Shenandoah Valley, and this rail line runs across the flattest and most central portion of the limestone valleys of Pennsylvania and Maryland. The Wilmington and Weldon Railroad was a lifeline to Lee's army at Petersburg. The route of this rail line extends almost directly north from Wilmington across the flat Coastal Plain of North Carolina. Around the middle of the state, near Goldsboro, it encroaches on the Fall Line, which runs diagonally across the state from southwest to northeast. Instead of continuing its northern path onto the Piedmont, the tracks begin to parallel the Fall Line, moving diagonally and staying always to the east on the softer strata of the Upper Coastal Plain.

All of these sedimentary challenges to railroads and roads were dealt with in dissimilar ways by different commanding officers. Some gained great advantage by being able to move armies across questionable terrain (Grant at Vicksburg), while others would end up losing their command when an army started sinking into the mud (Burnside at Falmouth). The next section of this text describes these challenges and explores what happens when mud is transformed into an even more hated particulate: dust.

PART III

Soft Rocks and Shovels

Conflict on the Coastal Plain

Is the mud knee-deep in valley and gorge?
What are you waiting for, tardy George?

> —Lyrics from the George Boker song "Tardy George," ridiculing
> George B. McClellan's averseness to engage the enemy in late 1862

Now we will rest the men and use the spade for their
protection until a new vein has been struck.

> —Ulysses Grant letter to George Meade after Cold Harbor

CHAPTER 9

A River Runs Through It
Flowing Water and Dissected Terrain

Stream Cuts and Strategy

Soft sedimentary rocks and unconsolidated sediments underlie all the battlefields from the Coastal Plain, including the numerous battlefields of the Peninsula Campaign, Cold Harbor, Champion Hill and Vicksburg, Bentonville and New Bern, and Shiloh and Corinth. These battlefields all have generally flat to gently rolling to undulating terrain, except where valleys have been carved by stream erosion. These fluvial depressions had all manner of impact on the fighting during the war. During the Seven Days Battles, for example, Robert E. Lee charged across these ravines at Beaver Dam Creek and Boatswain Creek (Gaines' Mill) before attacking parallel and between two stream valleys at Malvern Hill.

Most battlegrounds underlain by softer sedimentary rocks are part of a larger dendritic drainage basin. As such, the branching tributaries to a larger river will be oriented in a generally random direction, like the smaller branches on an oak tree. This pattern does not hold for rivers on the other side of the Fall Zone, where the harder crystalline and metamorphic rocks of the Piedmont have river systems than tend to trend from the northwest to the southeast—downhill away from the Blue Ridge and Valley and Ridge physiographic provinces. These rivers, including the Rappahannock, Rapidan, North and South Anna, and York, provided an inconvenient barrier for Federal forces wishing to move south toward the Confederate capital.

Unconsolidated Terrain and Tactics

West of the Fall Zone on the Piedmont the terrain tends to be rolling, with hard-rock ridges and domes providing tactically important high ground. For most of the Coastal Plain, in contrast, the landscape is generally flat, cut only by river valleys and the depressions created by the smaller fluvial tributaries. While the high ground of the Piedmont, Blue Ridge, and Valley and Ridge were repeatedly used for tactical advantage, the same can be said of the low ground, swamps, and stream channels of the Coastal Plain.

Rivers and streams of the Coastal Plain have a very low gradient, or slope, and most of their erosional energy is spent cutting horizontally, creating meanders, instead of downcutting vertically. The low gradient results in a lower velocity for the river when not at flood stage. This combination of meandering and lower water velocity produces broad, flat, swampy floodplains for many rivers. For some important rivers that cross the Fall Line, the term "river" is also misleading: the Rappahannock, James, York, and Potomac are all tidal as they enter the Coastal Plain, and thus they might be classified as estuaries or drown river valleys. From a tactical and strategic standpoint, these estuaries are wide and deep, with few fords available for rapid crossing. The rivers feeding the Chesapeake Bay across the Virginia landscape all flow from the northwest toward the southeast, providing at a minimum five fluvial obstacles that needed to be crossed between the two opposing capital cities.[1]

Field commanders on both sides used stream valleys and swamps as tactical defensive obstacles, allowing them to concentrate their forces on more favorable terrain; in other words, the officers could concentrate their forces in areas where they most anticipated an attack would occur.[2] A defensive line behind a stream or swampy ground would not need to be as strongly guarded, and the delay imposed on an attack provided by the water obstacle would allow time to bring up defensive reinforcements to bolster the line if necessary.[3] Streams and marshy ground also usually have little cover for attackers, meaning that an assault across such a landscape will be slowed on the killing ground just in front of a well-positioned defensive line. Floodplains are, after all, both plains and planes.

Examples of large rivers influencing strategy and battlefield tactics during the war are not hard to find (e.g., Burnside's debacle crossing the Rappahannock at Fredericksburg or Lee's vulnerability in retreat along the Potomac). A much smaller river and its tributaries, the Chickahominy, created chaos and disruption for both Union and Confederate forces during the Seven Days Battles, and this drainage system is the focus of the next section of the book.

The Chickahominy River and McClellan's Peninsula Campaign

The Chickahominy River flows for almost 100 miles in a direction roughly paralleling the York River to the north and the James River to the south. It joins the James near Jamestown and Williamsburg, on the south side of the Richmond-Yorktown peninsula. The orientation of the Chickahominy protects Richmond from an approach from the east: an army moving west (up) the peninsula from Yorktown will eventually need to cross the river to approach the outskirts of the Confederate capital.[4] The first 15 miles of the river flow across the Piedmont, but near Mechanicsville it crosses the Fall Zone and enters the Coastal Plain. On the Coastal Plain the river bends and winds across the landscape, and the river and many of its tributaries are surrounded by impassable swamps and flat, soft, muddy terrain.[5] The Chickahominy is an especially important river for the Confederacy, as it represents the last natural physical barrier east of Richmond, flowing south less than 12 miles from the city.

During glacial intervals of the Pleistocene epoch, the Chickahominy was actively incising into the underlying Neogene strata.[6] This left a deep river cut that would be backfilled during times of higher sea level. Muddy marsh sediments filled the incised valley, leaving behind a broad, swampy floodplain. The clay-rich soil and sediments left behind are impermeable, so that after the river floods, the water left behind will be trapped on the surface, creating boggy terrain that drains and dries at a very slow rate. The river varies in depth from under 10 feet northeast of Richmond to almost 70 feet where it joins the James, and river-bottom sediments are organic black and dark gray clay and silt, except where the river is adjacent to Neogene bluffs—then the bottom sediment shifts to brown and yellow sands.[7]

The discharge of the Chickahominy changes rapidly after heavy regional precipitation. The broad floodplain provides a misleading relationship between the stage (or water depth/elevation) of a river and its impassability. As the discharge rises, the stage only rises by a small amount; the river volume is spread across the floodplain, but the water level does not rise at a proportional level.[8] Nevertheless, even though the river is not getting drastically deeper, it is getting significantly more difficult to cross as the floodplain is turning into a wide, impenetrable bog. During severe flooding, it was often impossible to determine where exactly the river banks were, making the job of topographical engineers all the more difficult.

As McClellan's army approached the panicked city in the spring of 1862, the Chickahominy River would prove to be both a barrier and a boon. The initial Federal position in late May was bisected by the northwest-to-southeast–flowing Chickahominy and surrounding swamps; he had three corps north of the river, extending to Mechanicsville, and two corps south, near a crossroads called Seven Pines. The river barrier dividing his army was a seemingly minor inconvenience during dry weather; along the Federal line it was shallow and a mere 20 yards wide in most places. During poor weather, however, this 20-yard barrier became a mile-wide impasse when the adjacent floodplain and swamps were flooded. To the Army of the Potomac's great detriment, the weather was especially poor in mid- to late May, with water levels reaching a twenty-year high.[9]

To alleviate the fluvial threat to his communications and isolated corps, McClellan ordered construction of a dozen bridges across the river (see figure 9.1). His engineers, however, had a poor grasp on the surficial geology and alluvial deposits. They constructed bridges designed to span only the main river channel: when the waters began to rise the bridges needed to be extended a great length across the floodplain with corduroy roads. At maximum flood both the timber-clad road system and bridges were vulnerable to being swept away in the currents.[10]

On May 31, 1862, Joseph E. Johnston, commander of the Confederate defenses outside Richmond, decided to strike first. He intended to concentrate his attack on the isolated left wing of the Union army south of the Chickahominy. The Federal position was occupied by the Fourth Corps led by Brig. Gen. Erasmus Keys and the Third Corps of Maj. Gen. Samuel Heintzelman, with a line that extended from the White Oak Swamp to the south to the marshes flanking the swollen Chicka-

FIG. 9.1. This photograph was taken in June of 1862 by David B. Woodbury. This is one of more than a dozen bridges built across the Chickahominy by McClellan's engineers. Notice how still the water is when not at flood-stage. Enhanced detail of photograph from the Library of Congress.

hominy to the north. Johnston hoped to destroy these vulnerable corps before reinforcements could be brought across the river.

The night before the attack, meteorology, not geology, greatly benefited the Southerners. A typical, if somewhat nasty, thunderstorm dumped three inches of rain across Henrico County, flooding the rivers and local tributaries and further isolating the Federal left wing.

The Confederate assault got off to a successful, if tardy, start. Repeated, disjointed attacks had pushed the outnumbered Union units back almost a mile. After five hours of fighting the first Federal reinforcements, eight thousand men from Gen. Edwin Sumner's corps, began to filter into the rear of the Union position. It had taken these men hours to wade across the submerged bridges and floating corduroy timber roads across the Chickahominy, but their presence on the battlefield stalled the rebel surge until dusk fell.[11] About 7:00 in the evening the Confederate commanding officer was simultaneously struck by a musket ball and shell fragment, blasting him off his horse and leaving him unconscious on the field. Johnston's second in command, Gen. Gustavus Smith, was a civil engineer by trade. He surveyed a battlefield where the Union position had been dislodged eastward three miles to a third defensive line. Despite these setbacks, the final Federal position had been strengthened and was situated to take advantage of the swampy ground on both flanks and equally wet and poor terrain to their front. Nevertheless, Smith was determined to attack the Federal position the next day.

Early in the morning, three brigades under the command of D. H. Hill clashed with a larger force of Union infantry along the Richmond and York River Railroad. Harsh fighting continued along the railroad and in the surrounding woods for hours. Repeated and uncoordinated rebel attacks were successfully pushed back. The maneuvering and fighting was hindered by the thick forest; more than once infantry discovered they were firing on their own men.[12] Eventually, the Confederates pulled back, but the Union division under Joseph Hooker pursued and continued the fighting until mid-afternoon. The battle was over by 3:00 p.m. The Northern army had lost just over 5,000 men, with 790 killed. The Confederates had suffered a greater loss of more than 6,000 men, with almost 1,000 killed.[13] McClellan, who was tentative and cautious before the battle, became even more so after. He had his larger army construct elaborate earthworks in the deep, cohesive soil of the Upper Coastal Plain and sat idle waiting for reinforcements to be sent south. At the same time, Jefferson Davis was making a decision that would alter the strategic course of the war: the Confederacy's largest army guarding the capital would henceforth be commanded by Robert E. Lee.

In June both armies completed construction of extensive fieldworks running from the White Oak Swamp in the south across the swamps surrounding the Chickahominy River, terminating near Beaver Dam Creek and the town of Mechanicsville to the north. McClellan had more men, more field artillery, and importantly, more heavy siege artillery.

Lee convinced his president that a defensive stand in his trenches would only lead to a drawn-out and piecemeal defeat for his army in the face of McClellan's superior forces. He proposed to surprise and crush his unambitious opponent by launching an offensive that would exploit the river separating the armies. He planned to attack north of the Chickahominy, then drive southeast with the swamps and meanders of the river protecting his right flank. This battle plan contained one outstanding vulnerability: if McClellan acted boldly he would find Richmond lightly protected south of the river, and a quick strike toward the city would offer him a ten-to-one advantage in infantry power—but only if he attacked deliberately. Lee and Davis gambled that McClellan would continue his consistently cautious nature, and that the Federal

FIG. 9.2. Private Robert Knox Sneden's sketch map of Beaver Dam Creek clearly shows how Porter's Federals were aligned to take defensive advantage of the creek. A. P. Hill's men would need to cross this obstacle before encountering rows of abatis in front of the Union guns. Library of Congress.

commander would choose only to deal with the immediate threat to his right. They were correct.

Lee determined that Longstreet, A. P. Hill, and D. H. Hill should attack the Federal line behind Beaver Dam Creek closer to the Chickahominy River (west), while Jackson attacked the Union right flank farther from the river to the northeast (see figure 9.2). The Union position was strong for multiple interrelated fluvial reasons. First, the erosion of the meandering stream produced a three-foot-deep obstacle between floodplain swamps. On either side of the morass was flat terrain that needed to be traversed adjacent to the equally flat floodplain, and the Federal infantry and artillery were positioned on a gentle rise above the opposite creek bank. Even the low gradient of the stream and sweeping meanders proved beneficial to the defenders: any attack into a concave bend in the stream path left the approaching infantry vulnerable to enfilade fire on both sides from across the stream.

Thomas Jackson was late to the battlefield for the morning attack—a theme he would repeat throughout the Seven Days. By 3:00 in the afternoon, A. P. Hill had grown frustrated by the lack of communication from Jackson's command and decided to cross the Chickahominy in force. Within two hours his men had cleared the Federal presence in Mechanicsville and were in position across Beaver Dam Creek, facing toward the southeast. A. P. Hill now faced a critical decision: he had only a few hours of daylight left and his men were anxious for battle; however, he had still heard nothing from Jackson, who was supposed to be offering support by moving in strength against the Federal right wing. Hill decided to be ambitious.

Hill repeatedly attacked the strong Union line across Beaver Dam Creek, which proved to be both futile and costly. His men had to cross hundreds of yards of open ground under artillery and small arms fire with little cover before even reaching the creek. Those that made it across the stream were cut to pieces in the floodplain swamp in front of the Federal guns, positioned along several lines in depth. Even worse for the attacking soldiers, after their assault ground to a halt, they faced the deadly gauntlet of retreat when recrossing the open terrain between the creek and the protection offered by the buildings of Mechanicsville. Most soldiers simply sought shelter in the shallow swales along the creek, retreating only after the sun had set.

The first day of the Seven Days Battles had been a disjointed fiasco for Lee. Jackson never arrived on the field to offer support, and A. P. Hill had lost almost 1,500 men, four times as many as the Federals. When Jackson finally arrived north of Mechanicsville during the night, it made the Union line along Beaver Dam Creek untenable, and McClellan elected to construct a new line with better flank protection farther to the south. At the same time, he would switch his supply line and base from the north along the York River to the south on the James. The Army of the Potomac's new defensive line near Gaines' Mill would be positioned behind another tributary of the Chickahominy, Boatswain Creek. This position was critical to McClellan's army as the supply base was shifted between the rivers. The alignment of his troops also took advantage of the geology of the Coastal Plain. The Federal line was positioned on elevated older Tertiary strata behind a meandering creek, and muddy terrain lined both sides of the stream. The ground was easy to dig, and earthworks were quickly thrown up using the sand and silt.[14]

Lee struck again around noon on June 27 to begin the Battle of Gaines' Mill as A. P. Hill's division ran into Federal pickets north of the creek. This approach brought them into contact with Porter's division, arranged along three strong lines on the opposite side of the marshy creek. As Hill's men moved toward the Federal lines across the creek, a scenario similar to the previous day's fighting along Beaver Dam Creek began to develop: Confederate infantry were cut down by artillery and small arms fire as they crossed the flat plateau above the creek valley. The closest the rebels came to breaching the Union line was when North Carolinians of Dorsey Pender's and Lawrence Branch's brigades crossed the creek under heavy fire and attacked the Union line on the hilltop to the south. Unfortunately for Hill, these men were left largely unsupported and were forced to retreat in the face of strong defensive fire. Hill's attack was costly and accomplished little, and now his men were vulnerable to a strong Federal counterattack. To prohibit such a maneuver, Lee ordered additional attacks by Jackson and Longstreet along other segments of the creek. This prevented a Union strike in the sector of the battlefield that Lee thought was currently the weakest; however, the Union army launched an attack from a surprise location, using Slocum's newly arrived Sixth Corps, which had just crossed the river to support Porter. This strike stopped Longstreet's advance, driving the Southerners back across the creek.

As more Confederate reinforcements arrived on the battlefield, Lee distributed the brigades to launch a final broad assault across Boatswain Creek. The Southerners made slow and persistent progress in the face of heavy Federal fire, but at a high cost, until men of John Bell Hood's brigade finally pierced the Union line at the crest of the hill. Half of the brigade's men would be killed or wounded in the assault, but the rebels now had men behind the Union entrenchments and breastworks near the Watt House and the Union line began to disintegrate. Porter had already committed his reserve units earlier in the fight, so he had no way to stymie the Confederate gains, and soon his entire line was in retreat across the Chickahominy. Lee held the field as darkness came but had lost eight thousand men. Porter lost seven thousand men, which amounted to one-third of his entire corps, the largest in the Army of the Potomac.

After the battle McClellan was more convinced than ever that he was vastly outnumbered.[15] He had distributed his lines in the past to take full advantage of the impassable terrain offered by the Chickahominy and its tributaries and swamps. He would now go one step further, falling back to a position protected by fluvial obstacles on three sides. With his supply base now established at Harrison's Landing on the James River, Little Mac would have his army pass through the White Oak Swamp and form a new defensive line.

McClellan reasoned that, if Lee chose to continue his aggressive maneuvers, the limited paths of advance through the swamp would eliminate the advantage the Confederate general could gain from his presumably "superior" numerical advantage. Lee, meanwhile, hoped to continue the attack, trapping McClellan and the Army of the Potomac against a meander in the James but also always attacking far enough to the north so that the Union gunboats in the river would play no part in the eventual battle. As the Federal forces retreated south, McClellan's separated corps were attacked at a railroad depot named Savage's Station and along the northern border of White Oak Swamp at Frayser's Farm.

It was during the second battle that marshy terrain would truly hinder the Union army for the first time. Two roads became one as they ran south into the swamp, creating a bottleneck for the retreating Federal brigades, leaving them vulnerable to attack from the west. Jackson was poised to strike and crush the isolated Union rear guard in detail, but he seemed unable to find a way to cross the final tributaries and swampland in his path; the Federals had burned all bridges across the watery impasse as they retreated south. Jackson went so far as to wade into the swamp in desperate search of a way to move forward and expedite his attack.[16] Despite these efforts, the general reportedly dismissed reports from his scouts of fords available nearby and instead sat inexplicably motionless throughout the afternoon.[17]

Lee was frustrated by Jackson's lethargy and ordered Longstreet to attack late in the afternoon of June 30. The Union line would fall back again but hold against these assaults. The final defensive line occupied by the Union army was its strongest of the campaign. Malvern Hill was gently sloping with the Federal line traversing the highest available ground along the southern portion of the hill, 125 feet above sea level. The eastern and western flanks of the hill were protected by stream valleys, where the plateau-like terrain was abruptly ended by 50-foot lower

FIG. 9.3. Federal artillery position on Malvern Hill. Lee sent multiple waves of infantry across the fields below the cannon in a hopeless attempt to break the Union line.

FIG. 9.4 The discouraging Confederate perspective at Malvern Hill. To reach the Federal guns (horizon), Longstreet's men needed to cross this open field. No hope for a successful flanking maneuver was possible because of the local geology.

depressions.[18] Thus, the only available avenue of attack for Lee's army was across nearly a half mile of open ground, and this path rose 25 feet as it crossed over wheat fields and meadows (see figures 9.3 and 9.4).

As a result, Malvern Hill was an ideal defensive position with Neogene bluffs on the eastern side of the hill to provide a daunting challenge to attack, as well as lower, swampy ground to the east and northeast. McClellan ordered Porter's division—the same men who held the line along Beaver Dam Creek—to cover the front line of the Federal defensive position. The Union commander added the Third and Fifth Corps in support of Porter's men and kept Sumner's Second Corps in reserve. He also used the VI Corps to keep his line of retreat toward Harrison's Landing open. To make full use of the artillery firepower now available to him, Porter placed his field guns in front of his infantry lines, with his larger siege guns deployed a thousand yards to the rear.

Geologically, Malvern Hill was a strong defensive position and one that should not have been attacked. Even the gentle descent of the hill, with a slope of only 1 or 2 percent, was perfect for sending ricocheting solid shot and canister toward the approaching enemy. The plateau-like terrain toward the top of the hill also provided flat ground that allowed the gun crews to be fully supported by their caissons and teams, and the cliffs (and Turkey Run) to the west and Western Run and the swamps to the east assured the defenders from which direction they would receive an assault, allowing the artillery to be concentrated.

Lee would conduct a similar, and more famous, frontal assault over a similar landscape almost exactly a year to the day later. When he ordered Longstreet to attack the center of the Federal fishhook line on Cemetery Ridge at Gettysburg on July 3, 1863, he made his greatest tactical mistake of the war. Nevertheless, the assault on Malvern Hill was conducted on terrain that was even less favorable to offensive tactics. At Gettysburg, Pickett, Pettigrew, and Trimble attacked along a mile-long front against a defender on a slightly higher igneous ridge, and this ridge could not be entrenched in many places because of the thin regolith above the erosion-resistant igneous basement rock. The attack crossed a half mile of open ground. The curvature of the Federal position offered interior lines where reinforcements could be brought forward quickly to support any sector of the battlefield. Nevertheless, Lee always had the option of supporting Longstreet's assault with a renewed attack to the south (mirroring the previous afternoon's assaults) or a strike from the north out of the town proper. On Malvern Hill, in contrast, the geomorphology and topography assured that no such options were available because of the cliffs and swamps flanking the Union position. Thus, the Federals could concentrate their artillery and infantry lines facing directly into the heart of Lee's only viable path of attack.

In a battle plan for Malvern Hill that also mirrored that of the third day at Gettysburg, Lee brought forward his artillery to form a "Grand Battery" to bombard and diminish Porter's guns. In an unfair gun duel, the thirty-seven Federal pieces soon silenced the sixteen guns of the Confederate batteries. The Union artillery advantage was more than simply numerical: they had newer rifled pieces with better accuracy and range, all firing from a slightly elevated position. Longstreet brought forward additional guns, which were also quickly suppressed by the combination of Porter's field batteries and the larger siege guns to the rear of the Union lines, and naval gunfire from the river.

Around 4:00 in the afternoon, Longstreet ordered what was left of his artillery to retire, and the infantry was brought forward for the fateful attack. Lee ordered Lewis Armistead's brigade to lead the assault, with his supporting divisions using Armistead's advance as their signal to attack. Miscommunication immediately bungled the plan, as Armistead began his advance without waiting for Lee's direct command; the rest of the rebel army was not ready for the grand assault. Armistead advanced alone, and Federal artillery quickly drove his men to the ground. The infantry sought shelter behind a gentle swell in the ground that offered a small degree of protection in front of the Union guns. In an effort to support Armistead's trapped men as quickly as possible, a hodgepodge of brigades was thrown together in the woods at the base of the hill and ordered to attack. This assault soon withered in the face of Union artillery and rifle fire, and the only real threat to the Federals came as a sustained push by the Confederates toward the extreme left of the Union position by Mahone's and Wright's brigades. Federal reinforcements quickly thwarted this fleeting threat as well.

D. H. Hill's division was now brought into the fight as the previous wave of Confederate advances crested 300 yards in front of the Federal lines. Hill's men met a fate similar to Armistead's initial assault, as heavy defensive fire cut their ranks to pieces before the men hit the sandy ground in a gentle depression that offered at least a degree of protection from the Union guns. Here they remained until dark.

At this point in the battle the Federals were clearly assembling a terrific victory, and Lee decided to gamble on another assault, although this time he hoped that better organization and coordination would lead to success. This would be difficult, however, as the Confederate line was in chaos, and Longstreet was still sending brigades against the Federal left wing in a piecemeal fashion. At dusk, Lee ordered John Magruder and Lafayette McClaws to collect whatever rebel units were left and advance against Porter's line. Jackson would move forward in support on the left during this final assault. Devastatingly for Lee and the Confederate army, this assault proved as deadly as the previous ones, and any threat to the Union line was quickly squelched by Porter's reserve units.

The day had ended as a complete disaster for Lee and his army. In the previous battles of the Seven Days, Lee's subordinates often left him down (especially Thomas Jackson). During this final battle, blame for the failure of the army was more easily apportioned: Lee had created a battle plan ignorant of the strength of the Federal position, failed to recognize the geomorphological (terrain) limits on the tactics available during his aggressive offensive maneuvers, and lost command of his army and subordinates from the start. The Confederates suffered well over 60 percent of the eighty-five hundred casualties taken during this battle, men the Army of Northern Virginia would find difficult to replace.

McClellan was now faced with two logical (and appealing) options: he could counterattack against Lee's wounded army the next day or move his army west to imperil Richmond, compelling Lee and his distressed army to return to the defensive. Instead, to the chagrin of his corps commanders, he retreated 12 miles to the south to Harrison's Landing, a peninsula on the broadly meandering James River. Here he created an elaborate series of fieldworks on the river terraces and Tertiary lowlands around the north of his supply base, where he timidly waited to see what Lee would do next.

The Seven Days Battles had driven McClellan from the outskirts of Richmond, but at a cost to Lee of one-quarter of his army. The cost to the Army of Northern Virginia's leadership in the field had been especially concerning, as Lee had lost more than seventy-five subordinates, including ten brigade commanders. The Army of the Potomac had lost almost sixteen thousand men (with more than one-third of these missing and presumably captured), which was around five thousand fewer men than the Confederates.

Lee blamed his lack of "complete success," in his opinion, on a combination of swamps and a stone wall (Jackson). In his after-battle letter, the general made repeated references to the difficulty his men had with the tributaries and swamps of the Chickahominy: the Beaver Dam Creek provided the Federals a defensive position that was "a strong one" with "the banks of the creek in front being high and almost perpendicular." His attack fell short when "our troops forced their way into its banks, where their process was arrested by the nature of the stream." At Gaines' Mill he described how the Union infantry used the creek bank to hide sharpshooters and pickets, and he explained that his attack would need to deal with terrain that was "traversed by a sluggish stream, which converted the soil into a deep morass."[19] At Malvern Hill, Lee blamed the impassable swamps for limiting the volume of artillery he could bring into the fight, and the number of reserve brigades that could be brought forward. During the war Lee would have little positive to say of McClellan's skill as a general, but during this series of engagements he praised his opponent's ability to exploit the fluvial geology of the region:

> Under ordinary circumstances the Federal Army should have been destroyed. Its escape was due to the causes already stated. Prominent among these is the want of correct and timely information. This fact, attributable chiefly to the character of the country, enabled General McClellan skillfully to conceal his retreat and to add much to the obstructions with which nature had beset the way of our pursuing columns; but regret that more was not accomplished gives way to gratitude to the Sovereign Ruler of the Universe for the results achieved.[20]

During the Seven Days Battles McClellan had used tributaries of the Chickahominy to slow Lee's aggressive maneuvering while protecting his defensive lines and shielding his retreat. During several later battles similar creeks and stream valleys would prove to be a weak spot along otherwise formidable Federal defensive positions. In almost all cases these defects were formed by small streams that flowed perpendicularly to the defensive line, and thus the water was coursing across lower ground between and through the defensive positions, bisecting lines of entrenchments sited on higher ground.

The Army of the Potomac had taken defensive positions at Beaver Dam Creek and Boatswain Creek (Gaines' Mill) that were parallel to and behind the stream obstacles. As a result, the topographically low stream provided a water hazard to their front and created enfilade positions along the meanders: the streams became defensive force multipliers. Similar defensive lines were enhanced by fluvial systems at Gettysburg (Rock Creek), Sharpsburg (Lower/Burnside's Bridge across Antietam Creek), Fredericksburg (Rappahannock River), Sailor's Creek, and Stones River.[21]

Fluvial Defects to Defensive Lines

Twice during his Overland Campaign Ulysses Grant discovered, while on offensive maneuvers, a geological deficiency in his opponent's defensive position, and twice he tried to exploit the weak spot through a massed infantry assault. The first scenario, during the Battle of Spotsylvania Court House, led to temporary success; two months later the second maneuver was, Grant admitted, the worst decision of his military career.

During the early May 1864 Battle of Spotsylvania Court House, Grant launched two coordinated attacks against the center of the Confederate line. The easternmost attack, against the "Mule Shoe" salient, was directed by Col. Emory Upton, who employed an interesting tactic for his approach. Upton arranged his strike force with three columns of brigades, positioned four rows deep. This unusual configuration allowed the advancing men to present a smaller frontal target to the rebels in the salient, and at the same time, his men could exploit the geology of this portion of the field. Just to the front (northwest) of the salient, differential weathering between the Ta River Metamorphic Suite (more durable) and the rocks of the Quantico Synform (slightly less durable) created a swale that provided a defilade position for the attackers.[22] Upton's men used this protection when attacking and temporarily capturing the Confederate earthworks near the tip of the salient. Only a lack of ammunition and tactical support caused them to withdraw. The success of this advance through the geologic deficiency in the Southern line, if fleeting, prompted Grant to order a second, much larger assault at the same location. Instead of sending Upton's brigades, Grant selected Hancock's and Wright's corps for the next attack on the salient. This larger assault carried the Confederate lines, but Lee counterattacked and eventually slowed the Federal onslaught to the point that the rebels could construct a new series of earthworks across the base of the salient to the south. This reformed line held, producing a bloody stalemate by the end of May 12. Differential weathering created a problematic swale in front of Lee's earthworks, but this same weathering of feldspar-rich igneous and metamorphic rocks produced a thick, clay-rich soil that was ideal for the rapid construction of earthworks.

Two months later and fifty miles to the south, a similar military scenario would occur, but this time not because of weathering of metamorphic rock but fluvial geology. The first heavy combat of the Battle of Cold Harbor occurred on the afternoon of July 1, 1864, when Wright's Sixth Corps attacked Lee's newly excavated defensive line near the 1862 Gaines' Mill battlefield. The Second Connecticut Artillery, acting as infantry, found a poorly guarded portion of the Southern defenses where a small creek, later called Bloody Run, flowed perpendicularly through the entrenchments. The Federal unit was devastated by Confederate fire, but their reconnaissance in force created a gap that could be exploited by other Union infantry units who were moving forward. This attack surged into the fluvial weak spot, creating a temporary breach in the Confederate lines, and this success contributed to Grant's decision to launch a five-corps-strong attack two days later.

When Grant launched this massive assault on July 3, his advancing corps—fifty thousand men—again tried to exploit fluvial defects in the strong Confederate line, but the assaults were almost universally cut to pieces. Lee and his subordinates were learning the lesson about geological flaws in their otherwise formidable entrenched lines: trenches began to be constructed so as to be recurved at breaks along the line to cover and protect the ravines and stream valleys that flowed through the lines. When designing their earthworks, the Confederate engineers were clearly compensating for the tactical deficit created by the local fluvial geology.

The temporary breakthrough by the Union artillerymen was the only thing Grant had to show for his assaults at Cold Harbor, other than thirteen thousand casualties. He had not pushed the Army of Northern Virginia back into the Chickahominy River as he had hoped, and he had managed to inflict only five thousand casualties on his enemy. His goal in attack was not to capture Richmond or Petersburg but to destroy Lee's army's capacity to fight.[23] He had failed miserably. At Petersburg in the upcoming months, Grant refined his tactics, acting more cautiously after his army gained territory at the expense of a geologically weak portion of Lee's line. When he attacked a geomorphological deficiency or topographic low in the Army of Northern Virginia's entrenchments, it would be one created by his own army: the explosion of 7,500 pounds of black power under Elliot's Salient created a meteor-sized impact crater in the rebel line. Burnside's poor tactics in exploiting this breach led to yet another disastrous assault and resulted in the replacement of the corps commander.

Terraces, Bluffs, and Plateaus: Defending High, *Soft* Strata

High ground on battlefields is often the result of differential weathering of very hard rocks and slightly less durable rocks. Seminary and Cemetery Ridges at Gettysburg and Kennesaw Mountain and Pigeon Hill outside Atlanta are higher in elevation because they are composed of very hard and chemically resistant igneous and metamorphic rocks, and they are surrounded by slightly less durable (but still hard) sedimentary and metamorphic rocks. On many Civil War battlegrounds, the highest terrain is not created by differential weathering of hard rocks but by erosion of softer rocks caused by runoff. On the Coastal Plain, the same types of streams and rivers that troubled Jackson and Lee during the Seven Days created bluffs overlooking incised ravines, river terraces, or floodplains that were ideal defensive positions: the bluffs were relatively steep from past stream erosion, and many were 20 or more feet high, commanding swampy or flat ground to their front. Best of all, the soft ground on these heights could also be entrenched, unlike the high ground on the hard-rock battlefields. The next chapter will explore the exploitation of the erosion of this soft rock, and in doing so will document the failure of one commanding officer, Ambrose Burnside, to learn the folly of attacking well-entrenched defenders on high, soft sedimentary strata.

CHAPTER 10

Burnside and the Bluffs

Tertiary Bluffs at New Bern

On March 11, 1862, Gen. Ambrose Burnside and eleven thousand troops sailed from Roanoke Island, North Carolina, for the Neuse Estuary. Burnside's command was ordered to launch an offensive into North Carolina in support of McClellan's Peninsula Campaign, hopefully drawing Confederate attention and resources to the south. Rebel defenses along this portion of the Carolina coast were spread especially thin, and Burnside's men landed at Slocum's Creek without meeting resistance.[1] From here they marched north toward the town of New Bern, reaching the outskirts of the former state capital after a 15-mile slog. The next morning, they were planning on attacking the Confederate defenses south of New Bern, which were assembled across the Trent River. They would have limited support from their field artillery—poor weather meant that only the smallest guns had completed the sojourn north—but Federal gunboats on the Neuse could offer artillery support to their right.

The Confederate defensive line ran from Fort Thompson, overlooking the Neuse River on the left (east), across flat ground to a brickyard and railroad tracks; from here it doglegged to the north before continuing along bluffs overlooking a small creek called Bullen's Branch. This creek and the surrounding marshy ground created an obstacle from west to east along most of the right of the Confederate line, before the creek turned south near the railroad tracks.

The initial Federal attacks met with mixed results. Brig. Gen. John G. Foster's brigade moved toward the left of the Confederate position but was bombarded by friendly cannon fire from the river. Brig. Gen. Jesse L. Reno's brigade attacked toward the Confederate center at the brick-yard along the north-to-south–running rail line. As his men pushed the inexperienced Confederate militia out of their defensive line across a brick kiln, rebel cross fire halted their advance.

Union reinforcement moved forward into this weakened sector in the Southern defensive position, and the deficiency in the defensive plan-ning became more and more apparent. The Confederate line was rela-tively strong from the river to the tracks of the Atlantic and North Car-olina rail line. However, to continue the line directly to the east would have meant digging the entrenchments in or very near Bullen's Branch or near the base of the bluffs. The ground to the north of the creek and swamp was higher and much more easily defended, so it was decided to continue the line along the limestone bluffs. A proper defensive po-sition would have been fabricated in a relatively straight line, but Fort Thompson, the left anchor of the Southern line, had already been con-structed to the south. Thus, there was a substantial gap of almost 100 yards in the line where it shifted from south to north, and several redans were constructed to strengthen this weak spot (see figure 10.1).

The right half of the Confederate position was situated on the stron-gest defensive ground in the region. Parapets were constructed above the contact between the Oligocene River Bend Formation (a sandy limestone) and the Miocene/Pliocene Yorktown Formation (fossilifer-ous clay with fine sand).[2] As a result, the Confederate position ran from the river across the softer Yorktown Formation before continuing on the more durable River Bend Limestone to the north and west of the brick-yard. In short, the Confederate engineers had not taken advantage of the local geology when constructing the eastern (left) half of the line, but they had recognized the strength of continuing the line along the more durable and erosion-resistant bluffs marking the right extension of the line.

In addition to several lower redans built closer to the weak center of the Confederate line, eight other much stronger—and higher—redans were constructed to the west along the tips of elevated extensions of the limestone cliffs. These redans sat at the end of incised valleys that had been eroded into the limestone bluffs toward the north. The Confeder-ate line can be imagined as two giant, flattened hands, extending toward

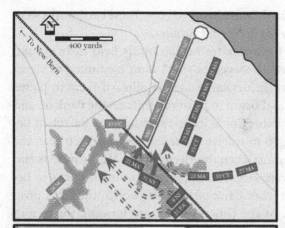

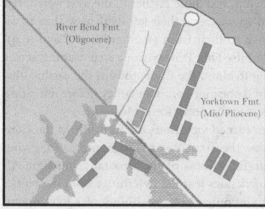

FIG. 10.1. Troop positions during the Battle of New Bern. Top: Federals (darker rectangles) gain success when breaking through Confederate center in the brickyard. This allows them to flank the stronger right of the Southern line located above Bullen's Branch. Bottom: The Confederate position was strongest where it was followed the bluffs created by the contact between the harder limestone of the Oligocene River Run Formation and the softer Yorktown Formation. The weak spot in the line, held only by militia, occurred where the line crossed the geologic contact.

the creeks to the south; the Confederates placed parapets across each fingernail, and the space between the fingers represents the low ground cut by the runoff of the incised valleys into the creeks to the south.

When Union reinforcements attacked again into the brickyard, they exploited the gap in the Confederate line and took maximum advantage of dead ground created by the eroded valleys where the creek turned toward the south (figure 10.1). The failure of the unseasoned Confederate militia in this sector of the defensive line allowed the Federal Third Brigade to move quickly through the low-lying, swampy ground adjacent to the railroad tracks and roll up the Confederate line toward Fort Thompson and the river. This maneuver widened the gap in the Confederate line until five companies of the Thirty-Seventh North Carolina were relocated from the left to halt the Federal advance. Concurrent with this action, Reno led the Union Second Brigade into Bullen's

Branch against the elevated Confederate redans. These earthworks were occupied by the Twenty-Sixth North Carolina.

For three hours the Confederates on the bluffs held their ground against repeated attacks. However, as the Union continued to strike against the brickyard, their inevitable success allowed them to pierce the Confederate center and begin to position units on the flank of, and even behind, the easternmost Confederate redans. When the rebels behind these parapets began to receive scattered fire from their rear, the situation became dangerous; when the Twenty-Fifth Massachusetts and Fourth Rhode Island were assembled along the railroad tracks to begin an attack along the weak flank of the earthworks, the rebel position became untenable. As the Union line moved forward, the men in the closest redans surrendered, and those farther to the west were faced with a second, new threat: an attack on their left flank as well as an assault across Bullen's Branch to their front (see figures 10.2 and 10.3). The Ninth New Jersey and Fifty-First Pennsylvania were wading across the cold creek and starting to climb the bluffs toward the earthworks. The rebels in the redans fired one more prudent volley before retreating rapidly into the dense forest to the north for protection.

The Battle of New Bern turned victorious for Burnside as the Confederates pulled back across the Trent River Bridge into town, giving the Federal force their largest success of the Coastal Campaign. Although defeated, the Confederates learned two things from the battle: first, they had mistakenly discounted the possibility of a joint operation against their coastal town. Fort Thompson, for example, had ten of its thirteen guns facing toward the river, leaving only three cannon to deal with an overland assault. Second, parapets constructed on Tertiary bluffs were an imposing defensive position, and they were made even more daunting to attack when a fluvial body like the three-foot-deep Bullen's Branch flowed in front of their base. Elevated terrain is valuable, but elevated terrain that can be excavated for earthworks is even more defendable. The only significant weakness to such earthworks occurred when they were attacked from the side or rear.

The Union army of eleven thousand men lost more men killed (90) and wounded (381) during the fighting but had fewer total casualties. The Confederate losses included 68 men killed and 116 wounded, but they had more than 400 men captured (or missing) after the battle. Burnside's force had only a single man missing. As a result, New Bern

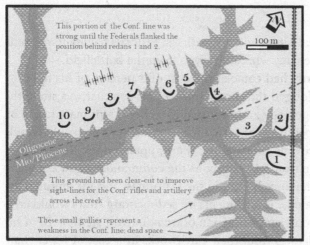

This portion of the Conf. line was strong until the Federals flanked the position behind redans 1 and 2

100 m

This ground had been clear-cut to improve sight-lines for the Conf. rifles and artillery across the creek

These small gullies represent a weakness in the Conf. line; dead space

Oligocene
Mio/Pliocene

FIG. 10.2. The Confederate position held by the 26th North Carolina overlooking Bullen's Branch was a strong one. Redans 4–8 are located on bluffs created by the contact between the River Run Formation and the Yorktown Formation. The weakest portion of the Southern line is located near Redans 1–3 and the brickyard, where dead space allowed Federal troops to move in force against the Confederate center and eventually assume enfilade positions towards both the left and right of the Confederate line.

FIG. 10.3. Left side of a Confederate redan occupied by the 26th North Carolina infantry. Note the lower, swampy ground in the background (Federal line of advance) that made these earthworks especially formidable. During the battle the trees below this position had been cut to improve sight-lines.

is considered a clear Union victory, with the Federals having fewer total casualties and an enemy who was driven from the battlefield.

Burnside's victory had come at the cost of 4 percent of his fighting force. He had dislodged the enemy from a position that was strengthened by the local geology and captured New Bern, a primary target of his army-navy expedition. Exactly nine months later he would attack another set of Neogene bluffs along the Rappahannock River in Virginia. This time he would be acting as the commander of an army ten times as large as the force he had in the Carolinas, and his repeated attacks against elevated entrenchments on sedimentary strata would cost him more than 10 percent of his 122,000-man army. At New Bern he drove the Confederates from their earthworks at a cost of fewer than five hundred men; at Fredericksburg he would fail to drive off Lee's army, at a cost that exceeded twelve thousand casualties.

Fiasco at Fredericksburg

Lincoln chose to replace the reluctant McClellan in the fall of 1862 as he envisioned a broad offensive against the armies of the south. McClellan had demonstrated a strange and frustrating combination of boastfulness in his communications with Washington and a timorousness in the field. The president elected to give the leadership of the Army of the Potomac to Ambrose Burnside, who was at the time the commander of the IX Corps. Upon promotion, Burnside immediately proposed a direct offensive against the Confederate capital along the Richmond, Fredericksburg, and Potomac rail line. Lincoln approved of this offensive maneuver, understanding that if Burnside could move quickly south across the Rappahannock River, he could force Lee to give battle on ground of his own choosing.

The planned route of advance would follow the Fall Zone contact between the Piedmont and Coastal Plain and pontoon bridges would arrive at the river prior to the army's advance. These would allow the Federals to cross the river rapidly, capturing the poorly guarded city of Fredericksburg. With the city quickly taken, Burnside would again move expeditiously to the south to threaten Richmond and force Lee to fight.

In mid-November the great army began moving, and within two days lead elements of the infantry were ready to cross the pontoon bridges and seize the city. Unfortunately, the pontoons were still in Washington, D.C., and the Union army would wait, frustratingly, for more than a week on the east side of the river overlooking the lightly guarded city. This delay, and Burnside's refusal to allow a portion of his army to ford the river upstream from the Fall Line, gave Lee the time he needed to recall Longstreet and Jackson from the Piedmont and Valley and Ridge provinces to the west. He distributed units from these Corps along 30 miles of the western bank of the Rappahannock to guard any likely crossing point for the Army of the Potomac. He also positioned Brig. Gen. William Barksdale's men in Fredericksburg proper to provide a warning in case Burnside acted boldly and tried to build his bridges directly below the city. On the night of December 11 this is exactly what the Union engineers attempted, and soon the initial construction efforts on the east bank of the river were under fire from rebel sharpshooters in the city.

For twelve hours the engineers were harassed by rifle fire from the riverfront buildings while the Union artillery on the Neogene rise called Stafford's Heights occasionally answered with a shelling of the city. Eventually, this balance between sniper fire and suppressive fire gave way to a new tactic: Federal infantry rowed several of the pontoon boats across the river to establish a bridgehead.[3]

This half-day delay allowed Lee time to design and construct one of the strongest, geologically enhanced defensive positions of the war. The city of Fredericksburg sits on a broad, flat set of river terraces.[4] South and west of the city run a series of rises that are underlain by older Cretaceous and Neogene formations. The key positions of Prospect Hill and Marye's Heights are both underlain by the Miocene Calvert Formation and Upper Pliocene Bacons Castle Formation (see figures 10.4 and 10.5). These geological units are clay-rich and more erosion resistant than the river terraces and Rappahannock floodplain. They overlook the Pleistocene terraces, which would witness nearly all of the fighting during the battle. This low-lying strata is composed of gravel, sand, and silt and has a generally flat slope (0.5 percent), with little available cover other than a few small scattered buildings.[5]

Lee's defensive line was fortified geologically in three different ways. First, and most importantly, the increased slope at the Pleistocene/Neogene contact provided an elevated position for his artillery and several lines of infantry. Jackson, to the south, had created an elaborate series of entrenchments along his defense in depth. Jackson's position overlooked a flat, half-mile-wide landscape toward the river. Lee's line snaked along the heights so that his well-sited artillery could provide

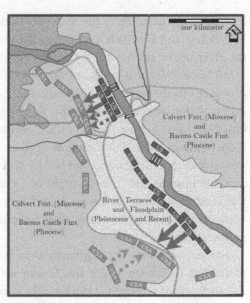

FIG. 10.4. After the Federal engineers constructed three sets of pontoon bridges to cross the Rappahannock, they attacked Jackson's Corps along Prospect Hill to the south and Longstreet's Corps on Marye's Heights to the north. More than a dozen assaults were made across the gently sloping terrain west of the city, and all ended in disaster.

FIG. 10.5. The famous stonewall below Marye's Heights at Fredericksburg. This photograph was taken on the 155th anniversary of the Federal attack against this strong defensive position. Confederate infantry was positioned along several elevated lines along this tiered slope, increasing the defensive firepower like a stacked casemate in a brick fort.

enfilade fire to support the adjacent sectors. Finally, the position was well entrenched with dug-in positions for both infantry and artillery. On many Civil War battlegrounds, the highest position—which is preferential for defense—cannot be further fortified by digging because of the shallow bedrock. This is true of battlefields underlain by igneous and metamorphic rock, like Kennesaw Mountain or Gettysburg, or softer (but still hard) sedimentary rock, like Stones River. At Fredericksburg, in contrast, the heights are underlain by sedimentary strata that are easy to displace and cohesive enough to require little in terms of support or revetting. Lee's men could easily and quickly dig in, and they were well concealed and protected during the subsequent Union onslaught.[6]

The battle began on December 13 with an attack across the southern portion of the battlefield by George Meade's division of Franklin's I Corps. This assault force of five thousand men began to move across the flat floodplain and river terrace against Jackson's line on Prospect Hill. The stubborn attack ground to a halt a thousand yards from the Confederate lines under intense artillery fire. Eventually the men began moving forward when counterbattery fire from across the river offered some relief, and Meade's men pushed toward the Richmond, Fredericksburg, and Potomac rail line and the woods marking Jackson's first line of defense. The intense hand-to-hand fighting along this sector of the line would mark the closest the Union army would come to piercing the Confederate line, but Meade's poorly supported units were driven back by a counterattack by Early's division. Jackson's defense in depth had proven successful.

During the initial delay in Meade's attack (when his men slowed, then halted in the face of heavy artillery fire during the late morning), Burnside commenced with a series of bewilderingly stupid assaults against the other wing of the Confederate army. Imagine and consider the scenario in which Burnside is giving directions to his subordinates prior to the attacks against Marye's Heights:

> *Picture Generals Sumner, French, Couch, Butterfield, and Kimball gathering around Burnside, in no way distracted by his facial hair, listening closely as their commanding officer begins to speak: "Gentlemen, I want you to capture two hills over there, across that vast, flat, open ground. Keep in mind that as you will make your approaches against the enemy you will need to cross an impassable drainage canal in your front that can only be negotiated on a few small*

bridges. This will almost certainly slow your advance and disrupt your lines. Oh, and you'll also have to attack one at a time—the ground to the north and south of here is too swampy for infantry to traverse, so our only viable path of attack limits us to deploying a few brigades at a time.

"By the way, the enemy has well dug-in guns on the hills and heavy artillery on that other hill down there as well. Seems like a terrific enfilade position against your lines. His infantry is entrenched along several lines on that bluff to your front. Oh, and I almost forgot to tell you—the closest of the enemy's defensive lines is also protected by a stone wall with a sunken road behind it. You can't really see the road from here because of the sod piled against it, so the face of the wall blends in perfectly with the surrounding terrain. Don't be surprised if the Rebels appear to spring up out of nowhere to cut your men down. One last note: If your first attack ends in complete disaster and your men are trapped on the field, you are authorized (here he winks . . . 'compelled') to repeat the offensive assault with new men, with absolutely no change in tactics . . . at least twelve more times. Let me know how things turn out. I'll be across the river in the mansion on the hill."

The attacks against Marye's Heights would cost Burnside in excess of five thousand men and hand Lee his greatest victory of the war. A month later sedimentary geology would contribute again to the end of Burnside's command of the Army of the Potomac: His famous "Mud March" bogged down, and the futility of this maneuver convinced Lincoln that he needed a new leader for his largest army.[7]

The strength of the defensive position represented by Marye's Heights is easy to underestimate when touring the modern battlefield. The flat terrain below the hills is today covered by residential neighborhoods and industrial parks, and the view toward Fredericksburg is obscured by a row of tall evergreen trees. As a result, the outlook from the Confederate artillery and infantry positions in and above the stone wall gives little perspective regarding the vulnerability of the massed Federal infantry surging toward the heights.

At New Bern, Burnside attacked a Confederate line that was sited on a topographic rise created by differential erosion between the Yorktown and River Bend Formations. This strong line of redans failed, not because of a direct assault from the front across a creek and swamp, but because the position was critically compromised when Federal troops pushed through the brickyard and into the flank and rear of the Confederate line. Similar strong positions were defeated when the defensive line was initially penetrated at Missionary Ridge, and at Spotsylvania Court House and Petersburg (at least temporarily).

At Fredericksburg, the well-entrenched Confederate line was never successfully breached and held by the Union army—the Federals never really got close. Along Marye's Heights, the Federal assaults failed to even reach the first of Longstreet's lines behind the stone wall. Burnside's apparent (if misleading) success in attacking a well dug-in enemy on Paleogene bluffs at New Bern likely contributed to his misjudgment of the strength of Lee's position on the Neogene bluffs at Fredericksburg. More questionable judgment by Burnside will be discussed in the next chapter, as the general sends his army marching through the mud in search of a repair for his reputation.

CHAPTER 11

Sediments and Morale

Mud and Dust

The vast majority of Civil War roads were unimproved dirt paths, often only one lane wide (see figure 11.1). When these roads entered swampy areas or when it rained, the compacted earth lost strength. Feet, hooves, and wagon wheels began to churn the sediment, decreasing the strength of the roadbed and increasing the softness of the soil, eventually turning the compacted soil of the road into mud. The roadbed's ability to support a load, whether the force was presented by the sole of a shoe or the wheel of a wagon, is its shear strength.[1] Sand and gravel rely on intergrain friction for their shear strength, while clay-rich dirt roads depend on compaction and cohesion between the clay particles.

From a military perspective, mud is simply a mixture of water and fine-grained particles of rock, mostly clay and silt.[2] The influence of this mud on transportation is dependent on the volume of water in the mud and the parent rock and grain size of the sediment. For example, the sandy roads across the Bentonville, North Carolina, battleground will drain quickly, but if the precipitation is persistent and the sand becomes saturated, the roadbed will lose all strength, prohibiting movement of any type. The roads around Vicksburg, Mississippi, lie atop loess, and the fine silt behaves in a similar manner to the North Carolina sand when disturbed.

Battlefields that have a higher clay content, such as Gettysburg, Chancellorsville, Antietam, or Petersburg, have roads that will become either sticky, with a little water added, or "mud holes," with much

FIG. 11.1. A Federal road construction crew from the 50th New York Engineers working across the North Anna river from Jerhico Mill. In the background the General Headquarters wagon train crosses a pontoon bridge while men bathe in the river. Library of Congress.

greater volumes of precipitation. During the Peninsula Campaign, Gen. Gustavus W. Smith reported on infantry advances that could not be accompanied by the desired artillery support "on account of the almost impassable conditions of the ground."[3] The degree to which this miserable wet sediment hindered movement is illuminated by W. H. Morgan in his memoirs: "No one who has not marched on foot behind any wagon or artillery train has any conception of what muddy roads are. Horses and mules were sometimes literally buried in the mud and left to perish, or shot dead on the spot."[4] Elisha Hunt Rhodes's exasperation with mud is clear from his diary entry for January 31, 1862: "Mud, mud . . . Will the mud never dry up so that the Army can move? I hope so, for I am tired and weary of mud and routine work."[5]

It can be argued that mud and clay even prolonged the length of the war, because Union general Ulysses Grant needed to wait until the mud created by persistent rains during the winter of 1864–65 dried and hardened before he could continue his aggressive maneuvering against Lee's battered army in southern Virginia. Seasonal muds were not the only sediment to influence logistics and even tactics. Mud found in coastal salt marshes slowed or prohibited more than one attack by soldiers of both armies. During the failed assault on Battery Lamar on James Island, South Carolina, the Third New Hampshire attempted a flanking maneuver that left them sinking into the soft, fluffy marsh muds within yards of the fort.[6] Trapped by the pluff sediment, the regiment withdrew along with the rest of the attacking Union infantry.

A similar situation occurred during the final battle of Fort Fisher by attacking Union infantry. Restricted in their attempted flanking movement on the western side against the land face of the colossal fort, the soldiers could not cross an expansive marsh, and their attack was compressed into a much narrower and more easily defended front.

A former U.S. Marine and history professor, C. E. Wood, wrote an entire book about the military history of mud.[7] He divided mud into three general categories: permanent mud (e.g., wetlands); seasonal mud (e.g., monsoon mud); and random mud (e.g., mud after a summer storm). The first two categories are relatively straightforward—and predictable—for military planners and logisticians, but random mud, from sudden storms or a rapid thaw, can create havoc for an army in the field, rendering movement on roads or across farm fields suddenly much more difficult.

The compressibility, or softness, of mud increases with the addition of water. This is a good thing for a soldier who needs to dig a foxhole or who suffers a fall from a great height, but sediment softness is the bane of transportation and mobility.[8] After soft mud on a road has been disturbed by wagons and marching men, more mud is created, an effect known as *churnability*.[9] The more churning of the muddy roadbed, the more mud that is produced, and the depth of the mud increases accordingly. The first units in a line of march will make the roads less passable for the units that follow, and the next units in column will continue the deterioration of the road condition.[10] This process continues until the road is impassable and the column of march becomes greatly extended.

Churnability was sometimes a problem for an army in pursuit. Grant commented on the conditions of the local roads left behind by his retreating enemy after Shiloh:

> After the rain of the night before and the frequent heavy rains for some days previous, the roads were almost impassable. The enemy carrying his artillery and supply trains over them in his retreat, made them still

worse for troops following. I wanted to pursue, but had not the heart to order the men who had fought so desperately for two days, lying in the mud and rain whenever not fighting.[11]

The viscosity of wet mud, or the fluid's resistance to flow, is directly related to water content and can slow marching soldiers and transport wagons in two ways. Low-viscosity mud will behave like a liquid, allowing wagons and cannon carriages to sink and become immovable. Decreasing the viscosity of the mud also increases the lubricity, so that horses and infantrymen will slip and slide. Higher-viscosity mud, in contrast, has a lower moisture content but becomes adhesive and cohesive.[12] With sticky mud soldiers will not sink as deeply into the sediment, but they will have trouble freeing their steps and invariably lose their shoes. James Crozer, of the Twenty-Sixth Iowa, described the cohesive effects of the Mississippi Valley with a note of exaggeration: "There is no bottom to the Mud here in Louisiana & it sticks like Glue. If one's boots were large enough they might take a whole plantation along with them."[13] Wagons will become stuck fast by the vacuum created as the mud sticks to the wheels. Even when freed from the morass the adhesive nature of the sediment will add weight to pants, shoes, or wheels, slowing all forms of progress.

The best dirt roads for travel with an army or wagon train were those that have compacted soil with gravel and sand for internal friction and a small amount of clay for cohesion, all in a solid state.[14] With the addition of water from a thaw or storm, a road passed into a semi-solid, then plastic state; with additional water added the sediment mixture reached, then crossed, the liquid limit, at which point a soldier was no longer marching but wading. If the process is reversed to the extreme—the roadbed completely dries and passes by the solid limit—a new potential hazard arises (so to speak) for the army: dust.

Dust consists of fine particles of solid material that become airborne. Because wind has much less competence than river water, the deposition of dust occurs in a very different manner than with fluvial sediments.[15] Two processes act together to produce dust. First, surface material must be pulverized and abraded by the application of a mechanical force (e.g., infantry marching in a column); second, the entrained fine-grained particles must be transported by air currents (e.g., the air disturbance caused by thousands of men walking).[16]

The loess at Vicksburg was the result of aeolian dust—fine silt produced by glacial erosion that was carried in great volume by wind across the dry, cold, Pleistocene Mississippi River Valley. The dust produced by an army column on the march is called "fugitive dust" because it "escapes" from the roads or fields, rather than coming from a particular point source.[17]

The composition of fugitive dust is dependent on the road material or ground cover; typical dust components include minute mineral fragments of silicon oxides, aluminum, calcium, or iron. The limestone Valley Turnpike would give off dust rich in finely crushed carbonate material when dry. Most of this sediment is extremely small, with about half the airborne content falling below 10 microns in diameter. As such, it can cause great discomfort to a soldier or pack animal, filling the mouth and choking the lungs. Col. B. F. Scribner of the Thirty-Eighth Indiana Volunteers described carbonate dust during his army's rapid march toward Louisville: "It was a rare and touching sight to see these poor fellows so covered with limestone dust that their garments, beards, hair and visages were all of the same color, all seemingly old and gray with the dust and bending under the burden of their guns and knapsacks, limping along with blistered feet."[18]

Dust could also form from road materials and rocks that were of a composition other than carbonate. Nurse Cornelia Hancock wrote of the dust of Virginia's Peninsula: "We have had scarcely one drop of rain for two weeks, the dust is shoe top deep."[19] One Federal infantryman wrote of the march, "One's mouth will be so full of dust that you do not want your teeth to touch each other."[20]

Once the dust is created, the spatial distribution of the cloud of material is a product of wind and grain size. Much of the 30 to 100 micron fraction will fall directly back onto the rear portions of the column, increasing fatigue and slowing the progress of the latter half of the line, so that the column of march becomes extended and strung out.

Great rising plumes of dust also give away the location of an army on maneuver. Contrast, for example, two marches by different corps of the Army of Northern Virginia: Longstreet's attempt to move around the Union left at Gettysburg was hindered by the bone-dry roads west of Seminary Ridge, while a month before, Jackson's famous surprise flanking march at Chancellorsville remained hidden because of the damp spring roads of the Piedmont. Longstreet's attack was unfortunately delayed, while Jackson's surprise attack was a battle-altering success.

A brief perusal of several dozen Civil War diaries from men fighting on both sides of the conflict indicates an almost identical number of complaints about mud and dust. Alexander Downing, of the Eleventh Iowa, for example, commented about how a hard rain could instantly cause a "deep dust to become a deep mud."[21] Dust and random mud slow or even prohibit the movement of an army, and there isn't much a military strategist or planner can do to foresee or avoid these hazards. Permanent and seasonal muds, on the other hand, are usually predictable and mostly avoidable, or at least mitigatable. There are, remarkably, many examples of field commanders neglecting to account for these types of muds, and one example above all stands out for the futility of the attempted maneuver.

Burnside's Mud March

Fresh from leading the Army of the Potomac to its greatest defeat of the war at the December 1862 Battle of Fredericksburg, Ambrose Burnside wanted, essentially, a do-over. Burnside proposed a plan to his superiors in Washington that would give him another crack at Lee at Fredericksburg, but this time on ground of his own choosing. He proposed to rapidly cross the Rappahannock with a surprise flanking movement to the west, after which he would march east, trapping the Army of Northern Virginia against the west bank of the river, where their formidable elevated entrenchments would be useless. Washington was hesitant, given that Burnside had proposed to rapidly cross the river and catch Lee off-guard a month before, only to then be delayed with disastrous results. Nevertheless, the prospect of having the much larger Army of the Potomac forcing Lee to give battle—with his escape route compromised—seemed too inviting to prohibit.

Early January 1863 had been relatively free of precipitation but cold in north-central Virginia, and the roads between Burnside's bivouacked army around Falmouth and Bank's Ford, eight miles to the west, were dry and mostly frozen. The divisions of Maj. Gens. Franklin and Hooker were ordered to begin the march on January 19; after they crossed a pontoon bridge at the ford, the grand division led by Gen. Edwin Sumner would follow.[22] The plan was to attack the Army of Northern Virginia on the 21st. Then it began to rain.

As the lead units started west, a nor'easter made its way into the

Mid-Atlantic. A typical low-pressure system might have slowed traffic and made men uncomfortable on the march. This system, in contrast, dumped more than three inches of cold rain across northern Virginia in forty-eight hours.[23]

The soil of this region of Virginia is especially poor for maneuvering in the rain. Most of the local soils are *ultisols*, clay-rich sediments that readily absorb water. Geographer Harold Winters described the soil's characteristics in detail in his article "The Battle That Was Never Fought: Weather and the Union Mud-March of January 1863": "It is commonplace for some of the uppermost and smallest particles in Ultisols to be transported 20 to 40 cm downward. This translocation results in a shallow argillic (clay-rich) horizon that, by clogging, limits penetration of water, thus favoring saturation of the upper soil during and after rainy periods. This may also result in standing water and encourage surface runoff."[24]

The continuous rain first lubricated the compacted clay of the roads, leading men—both mounted and not—to slip their way along the line of march. Soon the saturated roads began to be softened by the warming temperatures and deepening footsteps of the men to the point that churnability began to be a greater problem than slipping and sliding. Soldiers were sinking in the road up to their knees, and wagon wheels sank to the axel. William Swinton, a correspondent for the *New York Times*, reported at the time: "One might fancy some new geologic cataclysm had overtaken the world; and that he saw around him the elemental wrecks left by another Deluge. An indescribable chaos of pontoons, wagons and artillery encumbered the road down to the river. Horses and mules dropped down dead, exhausted with the effort to move their loads through the hideous medium."[25] As the rain continued and the men, wagons, and cannon persistently churned on, deep pockets formed in the sediment (see figure 11.2). The deepest of these could kill a mule.

Elisha H. Rhodes of the 2nd Rhode Island Infantry summarized the futility: "We found on leaving the woods that the roads were impassable by reason of mud. Daylight showed a strange scene. Men, Horses, Artillery, pontoons, and waggons were stuck in the mud. After making about two miles the waggons began to turn over and mules actually drown in the mud and water. . . . The mud was so deep that sixteen horses could not pull one gun."[26] Soon after the end of the campaign

FIG. 11.2. These three sketches by Alfred R. Waud were collectively titled "Why the Army of the Potomac Doesn't Move." This image was published by *Harper's Weekly* on February 22, 1862. Eleven months later the Army of the Potomac would be midway through the "Mud March." Library of Congress.

Gen. Alpheus Williams wrote about several such circumstances of sediment-induced animal cruelty:

> Tonight it looks like more rain. If it comes, the Lord help us, for I don't exactly see we can help ourselves. The roads will be literally impassable. In truth, they are so now. One can't go a mile without drowning mules in mud-holes. It is solemnly true that we lost mules in the middle of the road, sinking out of sight in the mud-holes. A few bubbles of air, a stirring of the watery mud, indicated the last expiring efforts of many a poor long-ears.[27]

Some units attempted to fashion corduroy roads across the muck, but even this usually reliable, if crude, solution failed.[28] Lt. John Mead Gould of the Tenth Maine described his regiment's efforts to battle the "pudding": "We came to a long reach, where all the rails, logs, and brush that we could find failed to make bottom for the wheels to rest on."[29]

On January 22 the rain stopped. Only one pontoon had made it to the now swollen river; the rest were stuck in the mud en route. Burnside called a halt to the march, instructing his men to stay in place as he decided his next move. His instructions were as clear as, well, a mixture of fine-grained sediment and water. As this indecisiveness was taking place, the cold winds began to blow from the north, chilling the air and preventing any significant degree of evaporation. With the element of surprise left sinking in the mud—Confederate soldiers were openly mocking the Federals in the slop from across the river—Burnside ordered his divisions back to Falmouth.

As his dispirited men began the long slog back to base, the local sedimentary geology found a different way to increase their misery. The well-churned muddy roads now lost enough moisture to make the sediment increasingly adhesive. Many men attempted to find their way home cross-country, leaving the roads entirely, but the intentionally churned (plowed) fields only increased infiltration by the precipitation and softened the mud.

It was not surprising, then, that morale was at an all-time low for the Army of the Potomac: more than two hundred soldiers were disappearing from the ranks daily.[30] Morale wasn't the only thing that was plunging; confidence in Ambrose Burnside, by his men and his superiors, bottomed out in the bottomless mud. By January 25 most men were back in camp, but Burnside was not. Lincoln had replaced him with Joseph Hooker as commander of the Army of the Potomac.

The infamous "Mud March" marked the final time an army would try a grand maneuver during the winter months in the Eastern Theater during the war. It was not the final time that mud created havoc on transportation or cost lives. Joseph Hooker's army had a similar defeat-in-battle/mud-march combination a few months later after the Battle of Chancellorsville and the heavy rains of April 15–30. At least his replacement, George Meade, would have good weather as he chased Lee north into Pennsylvania.

Fine-grain Sediments and Misery

Burnside's greatest mistake, with respect to the opinions of the men under his command, was to follow questionable tactical decisions in battle with questionable logistical decisions in mud. This combination of fighting and transportation disasters, when planning his flanking maneuver to cross the Rappahannock a second time, prematurely ended his short command. Subordinates who already doubted his skill as a combat general now had a growing sense of outrage at the futility of marching through mud. And mud doesn't limit its morale-diminishing effects to life on the road.[31] Elisha Hunt Rhodes summarized his feeling about the clay and silt well: "It is raining, and we all live in mud, sleep in mud, and almost eat in mud."[32] Rhodes leaves little doubt about his sentiment for the sediment.

The combination of mud and dust sometimes increased misery and lowered morale during the same campaign. Lee's Army of Virginia marched through dust in the Cumberland Valley to get to Gettysburg then retreated through mud to escape south across the Potomac. Grant's march on Vicksburg during the spring and summer of 1863 provides multiple examples of soldiers' complaints about both sedimentary hazards. The mud came first: William Wiley of the Seventy-Seventh Illinois wrote in his diary: "It was terable slipery and many a soldier got a fall in the mud that night. Stoped about 4 o'clock and laid in the mud until morning not knowing what minute we would be ordered forward."[33] Carlos Colby echoed a similar opinion of the mud:

> It rains here about half the time, the mud is so bad that it is almost an impossibility to haul anything on wagons. I have seen eighteen mules on one wagon, with not more half tons weight, most everything is carried by pack mules, they load them sometimes with two bales of hay, if Mr. donkey should be so unfortunate as to get mired, he is beaten for half an hour then unloaded and shoveled out, again to be reloaded.[34]

Two months later John Quincy Adams Campbell of the Fifth Iowa Volunteer Infantry wrote: "Our regiment was rear-guard of a division and as the dust was three inches deep, we got our share of Mississippi soil. Some of the boys thought they swallowed their 160 acres![35] If you ever travel all day on a road where the dust flew so thick that you could roll your tongue and spit out a good sized marble every eight or ten steps, you may have faint idea of the ordeal we had to pass through!"[36] The most significant difference between the two fine-grained hazards was related to the movement of an army and wagon train. Dust could make a march miserable but still tenable. Mud could kill momentum, movement, mules, and even wounded men. In short, no field officer ever lost his command after ordering a "dust march."

PART IV

Muddy Meanders
of the Mississippi

Those who do not know the conditions of the
mountains and forests, hazardous defiles, marshes and
swamps cannot conduct the march of an army.
    —Sun Tzu, *The Art of War*

. . . if my own regiment has not had a chance today to cover itself with glory it has with mud.
    —from the diary of Sergeant Osborn H. Oldroyd of McPherson's corps

CHAPTER 12

Geology of the Father of Waters

Fluvial Challenges

Of all the sedimentary battlegrounds from the Civil War, only those along the coastline are as dynamic as the sites along the Mississippi River. At Charleston and Wilmington, sand offered defenders an optimal construction material that minimized many Federal tactical and strategic advantages. It might be said that sand benefited the Confederates on a granular level. Along the mighty meanders of the Mississippi, the sedimentary geology influenced the fighting on a broader scale: sedimentary environs, in addition to sediment grain size, affected the maneuvers and combat in almost all conceivable manners. For the army, the twisting and winding river produced impassable bayous and highly defendable cliffs that needed to be captured.

Any and all engineering projects were constantly threatened by never-ending changes in the water level of the unpredictable river, and the possibility of a catastrophic flood diminished strategic boldness. Joint army/navy operations along the Atlantic coastline had to deal with predictable tides and an occasional storm or hurricane. Operation along the river had to deal with water levels that might fall a foot per day for two weeks, stranding vessels, or rise enough overnight to breach levies and inundate camps or drown engineering projects.

Despite these fluvial challenges, if General in Chief Winfield Scott's Anaconda Plan was to be accomplished, the Federals would need to capture New Orleans and the forts lining the bluffs above the many bends of the Mississippi. This chapter explores how the meanders, floodplains, deltas, and erosional scarps of the great river influenced the Union navy's and army's efforts to see the Father of Waters go again unvexed to the sea.

The Strategic Value of the Mississippi

The Mississippi River drains more than one million square miles of terrain and is navigable over most of its course; as a result, control of the river assured rapid transport for an army and supplies across a vast and critical swath of the Western Theater of the war. In the East, the opposing armies would bog down around Richmond. In the West, the armies needed to cross hundreds of miles of bogs and marshlands to strike at the enemy, and the easiest path to victory was on a boat. In short, the Western Theater had four times as much land to fight over compared to the East (as well as inferior roads and rail lines), making waterborne transport of men and supplies a logistical and tactical requirement.[1]

Capture of the river by the Union army and navy would essentially isolate Arkansas, Missouri, Louisiana, and Texas from the rest of the Confederacy, shutting down the transfer of critical food supplies from the Southwest. As the coastal blockade slowed the import of war matériel from European sources, more and more guns and supplies were finding their way to the Confederacy via Mexico and Texas. The capture of the river, and especially Vicksburg, would render this circuitous path of trade untenable.[2]

While the Mississippi River represented a significant barrier to east-west travel, it also provided an excellent corridor for north-south transportation, and the advent of steam propulsion meant that relatively rapid travel was possible both upstream and down. As a result, the capture and control of the river would allow midwestern agriculture to move once more by barge downriver and to the east. To establish this control over the river transport, the Union needed to capture the largest city in the south, New Orleans, and—by 1863—the best defended city in the Confederacy, Vicksburg.

The Federal Strategy

The national government pinpointed Vicksburg as a primary target from the start of the war. Early in the rebellion, flag officer David Glasgow Farragut sailed toward New Orleans with orders to capture the city and move immediately upriver to strike against the Mississippi stronghold. At the same time, the Mississippi River Squadron (or Union Brown-Water Naval Squadron) would move downriver from Cairo, Illinois, to eliminate key Confederate fortifications along the Cumberland, Tennessee, and northern Mississippi Rivers. This was, nevertheless, to be an army operation. The brown-water flotilla served a subsidiary role to the army, often acting as transport for infantry and artillery or as supply vessels. Maj. Gen. John Pope commanded the Army of the Mississippi in the field, and Ulysses Grant was in charge of the Army of the Tennessee. Maj. Gen. Henry Halleck was in overall command of both forces as he ran the Department of the Mississippi.

The Federal efforts to seize Vicksburg and control the river began in earnest in early 1862 with David Farragut striking upstream and Grant, Pope, and the brown-water fleet moving into Kentucky's and Tennessee's territory. In February Grant captured Forts Henry and Donelson, greatly bolstering his reputation.[3] In April he and the Army of the Tennessee were checked near the one-room log meetinghouse named Shiloh, just as Farragut was leading his naval fleet up the Mississippi birdfoot delta. Nevertheless, it would take Grant more than a year to make his way downriver to lay siege to the Gibraltar of the South.

The Geological History of the Mississippi River Basin

The only river in the United States that is longer than the Mississippi is one of the great river's own tributaries, the Missouri River. When the two rivers are considered jointly, their combined length of 3,710 miles rivals the Nile, Amazon, and Yangtze. Despite their nearly identical lengths, the Missouri River is dwarfed by the Mississippi when the discharge is considered: the Mississippi carries five times the volume of water and sediment as the Missouri.

The geological history of the Mississippi River basin begins approximately 800 million years ago with the breakup of the supercontinent Rodinia. During this rifting, the basin was fractured and stretched, leaving behind a deep valley known as the Mississippi Valley graben. When crust is stretched, as in a time of (super-)continental fracturing, faults will form that trend perpendicular to the extensional forces. Along these faults, large blocks of crust will sink, forming valleys known as grabens.[4] Surrounding blocks of crust on the other side of the fault may remain higher, forming seemingly—or later actually—uplifted terrain called horsts.

All of this rifting and faulting left the basin in a battered state, as geologists have construed it, a "geological battlefield." The Mississippi Valley graben is a failed rift system, with weakened, fractured crust that never fully separated during the subsequent 400 million years. Two more supercontinents, Panotia and Pangea, would accrete then break apart around the graben during this time; nevertheless, the valley would remain tectonically quiet over this expanse of time.[5] During the Cambrian and Ordovician periods, the future regions of Louisiana, Mississippi, and Alabama would lie beneath the Iapetus Ocean, and thousands of feet of carbonate rock (limestones and dolostones) would accumulate in their warm, shallow, marine waters. By the middle of the Paleozoic the Iapetus was closing, and clastic sedimentary rocks, composed of sand, silt, and clay, replaced the carbonate deposition. Erosion of the ancestral Appalachian Mountains brought massive quantities of sediment into the Mississippi River Valley, resulting in widespread subsidence from the mass represented by these transported sediments.

The Pennsylvanian period witnessed the assembly of Pangea and a collision with South America, forming the ancestral Ouachita and Frontal Ouachita Mountains in Arkansas and Mississippi and the southern ancestral Appalachians in Alabama. When Pangea later split in the Triassic, the southern Mississippi Valley was split as well, with rift basins extending from east to west across Louisiana, Mississippi, and Alabama. Twenty-five million years into this rifting, the Gulf of Mexico was born.[6]

The future Mississippi River Valley was now a giant sediment trap for the sediments shed from the uplifting and eroding Appalachian Mountains. The Atlantic coast had a similar trap in Maryland and Virginia, the Salisbury Embayment.[7] The sediments accumulating in this basin underlie battlefields including Fredericksburg, Petersburg, and the Seven Days.[8] To the west, ancient and southwestward-flowing rivers dumped sediment into the Mississippi Embayment. For more than 150 million years sediments accumulated in the basin, causing the ground to sag under the weight, and repeated rises and falls of sea level resulted in cycles of terrestrial and marine deposition.[9]

The transition from the Paleogene to the Neogene saw a general shift from marine and marginal-marine (delta) deposition to more terrestrial (fluvial) deposition. Accumulation of sediment in the embayment would extend from the Gulf of Mexico to southern Illinois and later erosion would leave a ring of Coastal Plain rocks and sediments that increase in age away from the axis of the embayment, which roughly coincides with the eventual path of the northern Mississippi River.

This 40,000+ feet of sediment represent a plunging syncline, a huge convex-down fold in the strata that dips along its axis toward the Gulf of Mexico.[10] Essentially, this collection of partially lithified rock and unconsolidated sediments resembles that of the Atlantic Coastal Plain, only bent into the shape of a giant horseshoe. Important battlefields like Shiloh and Corinth lie on the edge of this horseshoe.

Much more recently, from a geologic perspective, massive ice sheets advanced and retreated across the northern portion of the Mississippi River Valley. The most recent glacial advance twenty-two thousand years ago never covered the entirety of the valley; still, the drop in sea level and eventual meltwater runoff to the south profoundly shaped the regional topography. The surface of the ocean was fluctuating by as much as 450 feet, transgressing and regressing toward the north and south and altering the gradient of the Mississippi River.[11]

One of the more unusual sediments underlying some Civil War battlegrounds is the indirect product of glaciation. Windblown silt, or loess, collected after glaciers melted and retreated approximately twelve thousand years ago, and winds swept the sediment across the now dry floodplains.[12] These thick deposits are best seen on the downwind side of many large river valleys. Vicksburg, Mississippi, on the eastern side of the Mississippi River, is one such locality (see figure 12.1).

Loess produces massive (lacking layering or structure) cliffs when eroded by rivers or streams. These vertical faces are possible because of the thin film of calcium carbonate holding the silt particles together. This sedimentary characteristic was exploited by both armies, and even the citizens of the city, during the siege of Vicksburg when unsupported tunnels, and even entire dwellings, were easily carved into the subsurface.

During periods of regression, advancing glaciers picked up boulders and cobbles, grinding them along the barren landscape.[13] The result of this friction was massive quantities of tan or buff-colored silt. This fine sediment is known as glacial flour, and during glacial intervals the silt is easily winnowed from coarser sediment and carried across the cold, dry landscape by the wind, only to be deposited of the eastern side of the

FIG. 12.1. Silt-rich loess exposed on a 15-foot high cliff at Vicksburg, Mississippi. This sediment was especially good for mining operations because it was resistant to collapse and needed little in terms of bracing.

valley along the rising hills of the Coastal Plain.[14] Thick blankets of this silt collectively form a "rock" called loess, and the mid-continent "Last Glacial loess" is probably the thickest deposit of this type of sediment on the planet.[15] Vicksburg and Grand Gulf both lie on this blanket.

By around ten thousand years ago the Ohio and Mississippi Rivers had established the approximate configuration they follow today. The meanders of these rivers sculpt the wide floodplain that dominates the center of the valley, horizontally cutting meanders and cliffs into the mud and loess. The modern bird-foot delta didn't form until the last millennium, with the river shifting orientation and emptying into the Gulf of Mexico in five distinct geographic regions during the last eight thousand years.

Sedimentary Environments of the Mississippi River Valley

From Cairo, Illinois, to the south the Mississippi River is exclusively underlain by sedimentary rocks and sediments. These layers of rock thicken to the south, extending several miles underground until crystalline basement rock is encountered. At the surface, the Mississippi is incessantly transporting and rearranging unconsolidated sediments. The distribution and redistribution of this clay, silt, sand, and gravel by the water across the flat floodplain is the primary and almost sole agent responsible for creating bayous, natural levees, lakes, and erosional bluffs. For the most part these depositional environments and landscape features aided the Confederacy, providing high ground along the river for fortifications and bogs, swamps, and oxbow lakes that hindered nearly all hopes of rapid movement along the paths of approach for the Federals from the west, north, or east.[16] When the Union engineers attempted to take advantage of these geological features on a large scale, their efforts almost universally proved futile, often hindered by the shifting and flooding of the river itself.

The pebbles, sand, silt, and clay rearranged by the river in the floodplain presents an interesting heterogeneous concentration of sediment with different physical properties and engineering characteristics. The coarser clastics are more permeable, while layers of sediment rich in clay are more impermeable and cohesive. Without lithification, all of these detrital particles are easy to excavate, making canal construction an inviting tactic.[17]

In contrast to this mix of floodplain sediments, the eastern portion of the Mississippi River Valley is covered by a thick layer of loess. This light brown or pale yellow sediment is composed almost entirely of weakly cemented silt-sized particles of quartz and feldspar.[18] The tiny grains are angular and held together weakly by a thin cement, which is often calcareous, producing a slightly coherent layer of nonstratified aeolian "rock."

Loess deposits exhibit a property referred to as vertical cleavage, an important characteristic for citizens and military engineers alike at Vicksburg. As groundwater infiltration moves vertically through the silt, it produces planes of weakness in the direction of the water movement. Thus, most layers of loess will have a tendency to more commonly fracture along vertical faces instead of horizontal surfaces (see figure 12.2). As a result, natural cliffs are associated with thick loess deposits, creating a wall of sediment that might be excavated for a mine or, during exigent circumstances, a personal dwelling. Trenches cut into this type of soil will also need less revetting in the coherent, quasi-lithified sediment. In short, loess is an ideal sediment to conduct a siege, both above

FIG. 12.2. The loess below Vicksburg (literally). This photograph was taken of loess outcrops approximately 225 feet below Fort Hill. The light tan-colored sediment exposed in this roadcut on Washington Street (U.S. Bus. Route 61) holds a nearly vertical face despite apparent weathering and the creation of an over-steepened slope by the road construction.

and below ground. At Vicksburg, Grant would turn to this strategy after his multiple attempts to bypass the buff bluffs of the river continually failed.

The river's ability to pick up (entrain) and move (transport) sediment is controlled by the water's velocity. Faster-flowing water can entrain and move larger-sized particles, described as the river's "competence." A very fast-flowing stream or river may be able to carry gravel and coarse sand, while a slow, meandering river will be incapable of transporting anything larger than clay or perhaps silt; any larger sediments will settle to the riverbed. This, then, means that when water slows the competence decreases and sediment is dropped, or deposited.

A number of different factors can cause a river to slow and deposit sediment. When a river flows out of the mountains onto a flat plain, the gradient decreases dramatically, water velocity decreases, and competence decreases. The result is an *alluvial fan*. When a river flows into a larger body of slower-moving water, like a lake or ocean, the water slows rather suddenly, the competence decreases dramatically, and a *delta* is formed. When a river overflows its banks during a flood and spreads across the flat floodplain, there will be a continuum of velocities: the fastest flow will be in the center of the original channel, and the velocity will decrease across the floodplain toward each riverbank. Thus, the river velocity will begin to slow as the water flows over the floodplain near the original channel and will barely be moving along the edge of the river. Competence, in turn, will decrease away from the channel, across the floodplain, toward the periphery of the flood flow. Where velocity and competence decrease, sediment is dropped—the coarsest material will settle out on the floodplain closest to the channel (faster flow = sand), in the middle of the floodplain finer sediment will be deposited (medium flow = silt), and in the shallowest water near the riverbank only clay will be deposited. The result will be a pile of sand that runs parallel to the nonflooded river channel on the edge of the floodplain—a *natural levee*. These depositional features will later grow in subsequent flooding events and allow the river to grow deeper during heavy precipitation without necessarily flooding the entirety of the floodplain—until they are breached with catastrophic results.

Ulysses Grant would march his troops down the high ground on levees during multiple attempts to move south against Vicksburg. His engineers would also breach levees to create new waterborne avenues of approach to the rebel stronghold. His favorite tactic, however, was to bypass bends in rivers via canal, turning free-flowing meanders into lakes.

There is also a continuum in water velocity when a river flows around a bend. The water flows fastest on the outside of a meander, causing erosion and creating a cutbank, and slows toward the inside of the bend. Where water velocity decreases, competence decreases and sediment is deposited. As a result, sediment will accumulate on the inside of meanders, creating a depositional feature called a *point bar*. As the meander erodes on the outside of the bend and sediment is deposited as a growing point bar, the river channel will shift laterally, exaggerating the meandering of the river.

As a river grows more sinuous over time, two meanders may curve (erode) into proximity of each other. The river may cut off these loops (often when at flood stage), creating a shorter, more efficient course, abandoning one or both of the cutoff meanders.[19] These bypassed meanders are referred to as *oxbow lakes* because of their shape when viewed from above. Over time and with subsequent flooding, these lakes fill with clay and silt, and hundreds of former oxbow lakes can be identified across the Mississippi River floodplain. Ulysses Grant would also attempt to exploit these lacustrine depositional environments on more than one occasion when trying to send his troop ships south toward Vicksburg. In essence, his engineers were trying to reconnect the oxbow lakes with the Mississippi River channel, bypassing Confederate river batteries in the process.

The most interesting Civil War sedimentary deposit, from an "origins" perspective, is Island Number 10. This island, the tenth south of the confluence of the Ohio and Mississippi at Cairo, was created by uplift resulting from the most powerful earthquake to strike North America during historical times. In December 1811 and January 1812 the New Madrid region was jolted by a series of massive earthquakes resulting from movement along the deeply buried Reelfoot Fault system.[20] These shocks were powerful enough to be distinctly felt across the *Eastern* Theater of combat, causing damage to buildings in the District of Columbia and Richmond. On February 7 the region was struck again, and a 30-square-mile block of land was raised 20 feet above the surrounding terrain. This uplifted terrain, called the Tiptonville Horst, altered the flow of the Mississippi: the river changed course, deviating violently to the north and submerging the town of New Madrid.

With this dramatic uplift the Mississippi temporarily reversed course.[21] The river eventually reestablished the original, if altered, direction of flow to the south, but not without a great change in gradient and river velocity. Just before the river path encounters the Tiptonville uplift, the velocity of the river decreases and sediment, in the form of Island Number 10, was deposited.[22]

Each of these fluvial features and depositional environments affected the campaigns to capture and hold the river. The rest of this chapter focuses on these sedimentary influences on the conduct of military operations: while the Confederates established their strongholds on the cutbank vertical loess bluffs, Grant and Banks were determined to dig their way through the sediment to gain every possible advantage. The campaign would rage for two years across the mix of sand, silt, and clay of the floodplain before eventually ending in the most homogeneous sediment from any battleground from the Civil War, the silt under Vicksburg, Port Hudson, and Grand Gulf.

Meanders and Islands,
Earthquakes and Uplift

As Farragut was assembling his fleet to move against New Orleans and Grant was enjoying the success of his unconditional capture of Forts Henry and Donelson, Henry W. Halleck sent Maj. Gen. John Pope with eighteen thousand men to attack New Madrid, Missouri, and, just upriver, Island Number 10. Confederate forces had only recently abandoned Columbus, Kentucky, to take up position along a twisting, earthquake-induced double bend in the river. Both New Madrid and the fortifications overlooking Island Number 10 were positioned on the outside of consecutive meanders in the river, on opposite banks.

Halleck wanted Pope to strike rapidly overland at New Madrid, cutting off any escape route for the rebels. The swiftness of this movement was constrained, however, by the flooding river and the Great Mingo Swamp. The slow-moving army arrived outside the town on March 3 and began immediately to prepare for a siege. On March 13, after being supplied with large siege guns the day before, Pope began to bombard the Confederate positions. Under the fire of the heavy artillery, the town was abandoned just as quickly as it had been during the earthquake and flooding fifty years earlier. Within twenty-four hours, most of the men in New Madrid had crossed the river to reinforce Island Number 10, while others were sent downriver to Fort Pillow.

At this point, Pope's army and Foote's squadron of gunboats were separated by the fortifications on Island Number 10, and Pope had no way of crossing the river to strike at the rebels from the landward direction. Foote had commenced a bombardment of the island's batteries but with little apparent effect, and several of his boats had been damaged by return fire. Pope ordered Foote to run his flotilla past the fort to New Madrid to join him, but the flag officer resisted. He feared the firepower of the fortress and what it might do to his ships; a damaged or disabled ship was a captured ship because of the direction of the swift river current.[23] A disabled and captured ship might be repaired and used against his flotilla.

In the meantime, Pope's engineers were busy looking for other options to unite the infantry and the boats. Col. Josiah Bissell of the Missouri Engineers proposed to cut a canal across the base of the point bar opposite Island Number 10. The canal would begin just downriver from Island Number 8 and rejoin St. Johns Bayou and the Mississippi

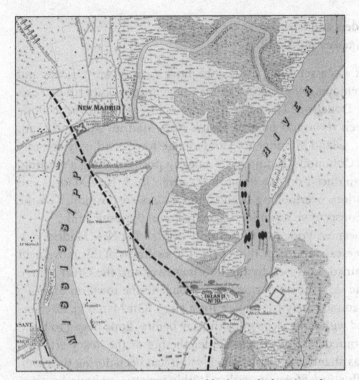

FIG. 12.3 This 1862 map by William Hoeckle shows the location of Island Number 10 as well as New Madrid. The canal used to bypass the Confederate fortifications runs across the top of the map. Added to this map is the approximate location of the Reelfoot Fault (dashed line). The Tiptonville Horst would extend from this fault to the west. Library of Congress. Fault location: VanArsdale, 2014.

River above New Madrid. Pope optimistically believed that once excavation was started on the canal, the natural erosive power of the river would do most of the dredging for his men.[24] The sappers cut through the natural levee at Phillip's Plantation, digging their way across the soft farm fields and into the woods for two miles. The sediment was easy to excavate, but the larger trees, and their roots, provided a formidable obstacle.

When completed the canal was 50 feet wide and 6 miles long; along its course it joined three bayous that needed to be cleared of trees (see figure 12.3).[25] Many of the previously cut trees also needed to be recut a second or third time as the river level fell.

By April 6 the canal was completely open. However, as the river level continued to fall, only flat-bottomed barges could negotiate the new

waterway. Despite these draft constraints, the canal allowed Pope to ob-
tain the barges necessary to cross the river and capture Tiptonville, the
horst's namesake, and the only road out of the Island Number 10 fortifications. This he immediately did.

Two days prior to Pope's crossing, several Union ironclads had successfully run the gauntlet through the Island Number 10 meander and
anchored downriver. Pope and Foote now controlled both ends of the
Tiptonville Horst, and Foote's gunboats could bombard the island from
both ends, which he immediately did. With the escape route blocked
for the Confederate garrison and the fortification suffering the effects of
flanking fire, the batteries surrendered.

The soft sediment of the floodplain, and the canal carved through
it, were certainly not a force multiplier for the Federals moving across
the floodplain. Nevertheless, the canal demonstrated a rather ingenious
way in which the local sedimentary geology could be exploited to gain
a tactical, or at least a logistical, benefit. Similar canal-digging projects were attempted later around Vicksburg, and in the Eastern Theater
along the James River, but neither would deliver any significant benefit
to their builders.

The surrender of Island Number 10 now opened the Mississippi to
Federal gunboats down to the next Confederate stronghold 60 miles to
the south, Fort Pillow. The Confederate batteries there were remarkably
similar in position and construction, from a geomorphologic perspective, to those at Island Number 10. Both forts occupied erosional bluffs
of loess above the cutbank of a meander, and both had a large accumulation of sand mid-channel below the guns on the hills. The primary difference between Number 10 and Pillow was the lack of batteries on the
sandbar of the latter. Fort Pillow was formidable, but the loss of Island
Number 10 and the fall of Corinth, Mississippi, necessitated evacuation
of the stronghold.

At this point, the time had come to concentrate on the next major
barrier to taking complete control of the Mississippi. Eventually, the
fighting would shift from sand, mud, and meanders to a different type
of material, one that was perfect for entrenching and defending: loess.

CHAPTER 13

The Vicksburg Campaign

Grant Does More with Loess

Bombarding the Bird-foot:
The Naval Strike Upriver

From the initiation of war, it was clear to both sides that Vicksburg, Mississippi, was the key to controlling transportation on the river. For the Northern government the initial challenge was finding a way to bring naval artillery and army infantry and siege guns to bear against the soon-to-be-fortified city. The terrain of the Mississippi River Valley made an overland campaign especially challenging, and any force hoping to capture Vicksburg or Grand Gulf would certainly need naval support: marching infantry and cavalry across Bayou Country is difficult enough, and moving heavy artillery by any means other than boat is next to impossible.

The Union strategy included capturing Vicksburg as quickly as possible via New Orleans, Baton Rouge, and Natchez. Flag officer David Glasgow Farragut was to lead an ocean-going naval squadron against the defenses of New Orleans and steam north past Vicksburg to join the squadron of gunboats making its way downriver from Cairo. This collection of shallow-draft river vessels was led by flag officer Charles Davis. The combined force of Farragut's and Davis's fleets would then work in conjunction to reduce the rapidly growing defenses around the city.

Farragut's fleet was poorly equipped for river operations, and from the very beginning of the campaign, sedimentary geology hindered his efforts, reduced his strength, and altered his tactics. Farragut assembled his fleet at Ship Island, a barrier island 12 miles off the coast of

Mississippi, and proceeded on March 7, 1862, to sail toward the bird-foot delta.[1] His fleet consisted of seagoing wooden vessels with deep drafts. In the Mississippi their screws would be vulnerable to damage from river-bottom obstacles, and their sails and rigging were essentially worthless. Without shallow-draft, flat-bottomed vessels, even entering the river channel would be a force-diminishing challenge.

Five navigable routes into the main body of the Mississippi were available to Farragut's fleet in 1862, and he chose to use one to the east and one to the west: Pass-a-Loutre and Southwest Pass. Prior to the blockade these two channels were dredged to provide deeper water for ships entering the main river channel. As the Mississippi enters the Gulf of Mexico the natural levees become submerged and sediment collects in a fan shape as a distributary mouth bar across the width of the river.[2] The combination of submerged levees and the mouth bar, when not dredged, limits the depth of the water and, as a result, the maximum draft of any ship attempting to enter the river channel through the delta.

The smaller vessels, including a flotilla of mortar boats, passed easily over the bar at Pass-a-Loutre. Farragut's larger, deeper-draft ships attempted to enter the river at the larger Southwest Pass, with mixed success. Not only was this a more indirect route upriver, the deeper-draft ships repeatedly ran aground on the sandy distributary bar. It took until April 8 for most of the ships to cross the bar and enter the main river channel; unfortunately for the Federal fleet, Farragut's largest ship, the *Colorado*, was left behind in the Gulf. She simply drew too much water at 22 feet, and the task force moved upriver without her firepower. Sedimentary geology had already reduced the weapons available to Farragut, even if by only twenty-five guns.[3]

Farragut wasn't a lawyer, but he did eventually pass this bar, and the Federal ships moved north across the bird-foot delta with 166 cannon and 26 howitzers. They were joined by a mortar-boat flotilla commanded by David Dixon Porter, adding twenty-two smaller ships to the force. The primary protection afforded to the city of New Orleans came from two stout forts located 75 miles downriver from the city waterfront. Fort Jackson was the more formidable of these brick fortifications, mounting 75 cannon. Across the river sat Fort St. Philip. This smaller fort, and a newly added water battery, added fifty-three guns to the river defenses. Unfortunately for the Southerners, half of these guns were older 24-pounders.[4] The defense of the pair of forts was enhanced by the twisting of the river, according to naval historian Fletcher Platt: "The main feature of the combination was that any ship coming up would have to slow for a sharp right-angled turn under the artillery, just where the current runs swiftest."[5]

The Federal plan to strike at New Orleans called for cooperation between the army and navy. Porter's mortars were to bombard the rebel forts while Benjamin Butler led eighteen thousand infantrymen across the delta on foot from Isle au Breton Sound to threaten the forts from a landward direction. This would leave Farragut an opening to run past the distracted and hopefully battered forts to travel upriver and capture the (largely) undefended city.

By April 17 Porter's mortar boats were within range of the forts. They were positioned to take advantage of the geomorphology of the winding Mississippi: instead of anchoring directly downriver within view of the target, Porter halted before rounding the final meander below the forts. This left two-thirds of his ships hidden behind the point bar deposit but still within gun range of the forts. The high arcs of his mortars could fire over the scrub- and tree-covered point bar, using indirect fire against the forts while remaining hidden. Farragut anchored his own fleet a short distance downriver from Porter's flotilla, waiting for the mortar boats to conduct their business.

On April 18 the mortars commenced firing on the stronger of the two Confederate citadels. By the end of the day, more than a thousand 13-inch rounds had fallen on or around the fort, although most of Fort Jackson's guns remained operational. Firing began anew early on April 19, and this time more of the shells found the parapets and gun platforms; nevertheless, Confederate casualties remained light and the fort persisted as a defensive threat.

Over the next three days Porter's bombardment would become less and less effective, in part because of the muddy sedimentary geology around the forts. His exhausted gun crews were spending less time precisely aiming their weapons, and their inconsistent timing fuses were often causing the shells to explode mid-flight. As a result, the crews were instructed to set the fuses to explode after the shells found "solid" earth. This change in bombardment tactics further diminished the effectiveness of the shellfire—the rounds were burying themselves in the soft Mississippi mud, sending only a shower of silt and clay skyward when, or rather if, they detonated. All shrapnel was defeated before it reached the surface, rendering anything but a direct hit a mere nuisance.

On the night of April 20, Farragut ordered two gunboats upriver to sever a thick chain the Confederates had extended across the river. With their success, only the final barrier to New Orleans, the gauntlet between the forts, remained. Farragut divided his squadron into thirds, with the first eight vessels moving upriver where they were to concentrate suppressive fire on Fort St. Philip. Farragut would lead the next set of vessels from his flagship *Hartford* to pound Fort Jackson. The final third of his squadron would provide fire support against whichever fort was proving to be the most trouble.

Sediment was used in two unusual ways during the final preparation for the raid upriver: dark brown adhesive mud was spread on the hulls of the ships to reduce nighttime visibility of the vessels, and sand was spread on the decks to provide traction and absorb soon-to-be-spilled blood.[6]

In the middle of the moonless night on April 24 the fleet began moving upriver. A little after 3:00 a.m. the Confederate sentries spotted the lead vessels of the flotilla, and great bonfires were lit on the riverbanks. Large rafts were also set afire and sent drifting downriver.

Confederate artillery fire did some damage to the Federal ships, but in the still night air the gun smoke of the cannon fire soon reduced visibility and made aiming accurately more difficult. Farragut's *Hartford* ran aground before the guns of Fort St. Philip just as a new threat arose for the Federal flotilla: upriver from the forts lay eleven gunboats of the Confederate River Defense Fleet.

A series of violent clashes between individual ships took place in the swirling waters around the forts with several vessels being rammed and badly damaged on both sides. Friendly fire from the forts added to the confusion. Three hours after the ship-to-ship fighting began, the River Defense Fleet was no more. The Federals suffered just under two hundred casualties and lost one ship, the 1,300-ton gunboat *Varuna*. Nevertheless, Farragut was now north of the forts.

The next afternoon, April 24, Farragut began the journey upriver to the city. The slower mortar boats were left behind to deal with the still-dangerous forts. Porter was to demand surrender of the forts and, if refused, restart his bombardment. In a worst-case scenario, the mortar flotilla would be used to support Butler's still-in-route infantry during a later siege.

On the morning of April 25 Farragut's fleet steamed past the last remaining Confederate batteries along the river below the city and anchored with their guns overlooking City Hall. The levees raised the elevation of the river relative to the sunken city and, in doing so, provided a superior firing position for the Union gunboats. From this threatening position, Farragut began surrender negotiations with city officials. His bargaining position was favorable but not perfect: Forts Jackson and St. Philip remained in Confederate control.

On April 27 Butler's troops began their amphibious invasion. Faced with the dual land and water threats, the Confederate commander of the fortifications considered surrender. When Fort Jackson's garrison mutinied, the decision was taken out of his hands. Butler's men occupied both forts by the end of the month. Now, with the forts in Union possession, "The Big Easy" matched its nickname and finally surrendered. Farragut had accomplished a terrific victory at little cost and demonstrated a new tactic that the Union navy would repeat in the near future: when faced with a strong fortification, sometimes the best approach is to avoid a direct confrontation and bypass the threat. After all, whether permanent or temporary, fortifications lose much of their strategic value when fighting (or transportation) moves elsewhere.

With the fall of New Orleans, Confederate efforts to improve the defenses of Vicksburg increased rapidly. Farragut, meanwhile, took pause to repair his ships, rest his men, and resupply the fleet. His orders were to steam upriver at once, run past Vicksburg, and join Flag Officer Charles Davis's river gunboats. From here the combined gunpower of the two flotillas could be hurriedly brought against the immature Vicksburg defenses from the north and the city could be captured.

When Farragut eventually began his sortie upriver he found his ships—all oceangoing vessels—ill suited for river travel. His ships sat deep in the water and seemed to be constantly running aground. Moving against the river's current also burned an excessive amount of coal, so the vessels needed near-constant resupply. Making the slog north even more intolerable, dysentery and malaria were a growing problem for the sailors and infantry that had been brought aboard as an eventual occupation force.

Farragut captured the undefended city of Baton Rouge without incident and continued upstream past Natchez to Vicksburg. By May 18 his ships were anchored downriver from the city, hidden behind the final river meander to the south. From here he sent the sloop *Oneida* north to demand surrender of the city and her defensive works. The demand was met with derision by the Mississippians, and for good reason.

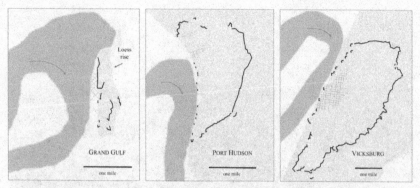

FIG. 13.1 The Confederate defenses at Grand Gulf, Port Hudson, and Vicksburg were all constructed on loess hills on the outside of meanders in the Mississippi. From this position on the heights of the cut-bank, their batteries commanded a large stretch of the river upstream and down, assuring that they could commence firing before Union ironclads and steamers could bring the full weight of their broadsides to bear.

Farragut was hardly in a position to negotiate from strength. In fact, his ships were in a terrible strategic position, in large part because of the local fluvial geomorphology. The mighty river meanders did much to diminish the Federal fleet's firepower. The vast majority of the Union guns on these seagoing vessels could only be fired in broadside: a direction roughly perpendicular to the axis of the ship. The river's meanders meant that the guns could not be trained to fire at a target upriver while the ship was moving against the current (see figure 13.1). This meant that the majority of Farragut's ships would not be able to fire with full effect until they were adjacent to Vicksburg's batteries, and they would predictably suffer damage steaming against the current to get into this firing position.

The cutbank of the Mississippi meander also eroded high bluffs into the loess below and west of the city. The Confederate guns were thus in a position high above the river, at a height that Farragut's cannon couldn't always reach. Seagoing artillery is designed to fire at distant targets that are on the horizon, not 200 feet above sea level. As a result, the naval guns couldn't be elevated to the degree that would allow them to hit rebel batteries on the loess bluffs (see figure 13.2).

Without a means of effectively silencing the city's elevated batteries, the occupation force of approximately 1,500 infantry was useless. Faced with these challenges, and much to the anger of Washington, the Federal fleet commander decided to take a conservative approach. Instead of attacking or running past the city, he ordered the anchors to be

FIG. 13.2 This photograph was taken looking west from the Fort Hill, the highest Confederate strong-point along the bluffs overlooking the Mississippi River. The waterway coursing across the flat floodplain is not the Mississippi, which bypassed northern Vicksburg via meander cut-off in 1876, but the Yazoo Diversion Canal. This photograph was taken from an elevation of 337 feet; the floodplain below is 60 feet above sea level.

raised and his ships to turn around. The fleet returned to New Orleans after completing what Farragut now conveniently termed a "reconnaissance in force."[7]

Had he been able to run past the Vicksburg batteries, Farragut would still have had to travel far upriver to join Davis's flotilla. The Confederates had assembled a small fleet of rams that were harassing the brown-water fleet. On May 10 the rebels attacked near Fort Pillow with success before retreating downriver to Memphis. The Confederate rams, however, were decimated after Davis acquired nine rams of his own and, on June 6, attacked with a combined gunboat/ram squadron. That same day Farragut began steaming upriver again, after being severely chastised by the president for his previous apparent lack of determination.

Two weeks later Farragut arrived for a second time below Vicksburg, with a much-improved force: his transports carried more than three thousand infantry, and he had added a fleet of mortar schooners. As he had done at Forts St. Philip and Jackson, Farragut ordered his mortars to bombard the Vicksburg defenses. This they did for two days, raining

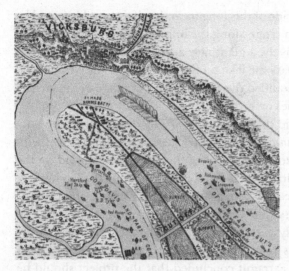

FIG. 13.3. This detail of an 1863 map by L. A. Wrotnowski illustrates well how the canal (bottom center) would bypass the majority of the batteries protecting Vicksburg. Note, however, that the downstream (right) portion of the canal is still within artillery range of the southernmost Confederate guns (above the point of the arrow). Library of Congress.

shells down on the Vicksburg batteries, and Farragut began to sail up-river in two columns, sans the slow mortar schooners. Once alongside the bluffs of Vicksburg, they opened fire, throwing the majority of their shells into the loess cliff faces. Only the broadsides of the flagship *Hartford* seemed to reach the Confederate batteries sitting high above the water.[8]

The Confederates, in return, sent deadly plunging fire toward the vulnerable wooden decks of the ships below. As rapidly as possible the Union vessels steamed north through the gauntlet, and after escaping the maelstrom of the Confederate artillery, Farragut found that eight of his eleven ships remained in formation. He had bypassed the city but lost forty-five men in the process.[9]

The Federal soldiers were disembarked upriver of the city on the Louisiana bank. For the next month these men, led by Brig. Gen. Thomas Williams, would undertake an engineering project aimed at bypassing the guns of Vicksburg, rendering the fortress pointless (see figure 13.3). According to Williams, "the Mississippi will take the course of the cut off and Vicksburg becomes an inland town with a mere creek in front of it. So the batteries will be made useless, and Vicksburg will fall with the spade."[10]

In a tactic that would be attempted repeatedly during the next two years of the war, men began digging a canal to cut across a narrow portion of the point bar of the Mississippi River meander. Vicksburg sat on the cutbank on the outside of this long meander; if a point bar could be traversed out of cannon range along the opposite bank, the city would be sitting above, effectively, a giant new artificial oxbow lake. Federal ships could simply bypass the lake on the new path of the Mississippi, staying in the freshly modified river channel out of range of the menacing Confederate batteries on the hills.[11]

Farragut would spend the second half of July dealing with the 800-ton rebel ironclad *Arkansas*. This gunboat had surprised his fleet with an attack out of the Yazoo River. During this embarrassing raid for the Union, Farragut lost thirty men and had every wooden vessel in his fleet damaged to some degree, all from a single, solitary enemy ship.

To make matters worse, the canal building was progressing at a glacial pace. Plagued by dysentery and malaria, the men grew weaker by the day. Gen. Williams requested the men be transported to Baton Rouge to recover, and Farragut concluded that the project should be abandoned. Within a week Farragut, who was concerned with the dropping river level in late summer, began to steam downriver once again. Vicksburg had successfully survived the first two Federal efforts to capture and/or bypass the city.

On October 25, 1862, the Department of the Tennessee gained an aggressive new commanding officer: Ulysses S. Grant. Grant immediately proposed to move his forces south through Tennessee and Mississippi, taking a route that paralleled the river but ran 60 or so miles to the east. His target was the capital of the state, the railroad junction of Jackson. From here he would strike west and take Vicksburg from the rear. At the same time, Maj. Gen. John Alexander McClernand, after much politicking and pleading, was given permission by Washington to move down the river directly to make an amphibious assault against the city. This move both puzzled and frustrated Grant, who was in overall "charge" of the forces in the vicinity of the river, and he determined to launch his own preemptive strike down the river. Grant ordered Maj. Gen. William Sherman and his thirty thousand men to be escorted down the Mississippi on transport ships to the mouth of the Yazoo River by Adm. David Porter's gunboats. From here they were directed to attack the Confederate positions on Chickasaw Bluffs.

The Chickasaw Bluffs lie adjacent to the Mississippi River floodplain. This sudden increase in elevation, ranging from 50 to 200 feet, represents an erosional contact where the former Mississippi River's path encountered loess hills. Sherman's men faced a daunting task: they were

going to have to cross muddy and boggy floodplain terrain before attacking uphill against the Confederate earthworks on the loess rise.

By the end of December, the Confederates had around twenty thousand men on or near the bluffs. Sherman probed slowly toward the position, constantly delayed by impassable ground and water barriers. On
December 28 his men were ready to cross from the floodplain onto the
higher loess topography. The Confederates were ready: to reach their
line, the Yankees would need to cross a bayou, more muddy terrain,
and a line of abatis. Only a single causeway crossed the bayou, limiting
Sherman's initial striking power. Even in the plan of attack, a relic of
the ancient meanders appears to have acted against the Federal infantry. The commanding officer of the Forty-Second Ohio, Capt. William
Olds, reported: "The proposed point of attack upon the bluffs proved
to be the interior arc or semicircle, so as the storming Brigade advanced
it found itself in the center of a converging fire."[12] Brigades sent into the
fight to support these men became mired in mud or lost along the curving levees of the swamp.

The silt-rich loess has the ability to hold near vertical cliffs, and the
Confederate line used this attribute to gain outstanding firing positions
at the edges of the bluffs. The terrain, and the poorly conceived battle plan, resulted in a debacle for Sherman. He lost almost two thousand men, and the Confederate casualty list was an order of magnitude
smaller. Despite this setback, Sherman intended to renew the attack at
the location upriver, but heavy rains and a rising river convinced him to
withdraw.

Sherman's men were handed over to McClernand after the Chickasaw Bluffs disaster, and their new leader followed Sherman's sugges-
tion to move up the Arkansas River and strike against the isolated Fort
Hindman. This they did, using a joint operation with added firepower
from Porter's gunboats. On January 11, after a valiant if futile defense,
the fort surrendered.

McClernand's victory did little to promote his position in the command structure, however. Grant did not trust his subordinate officer
and rival, considering him a political general who had stepped out of
bounds when approaching Washington with his own river-borne attack plans against Vicksburg. Grant decided the best manner in which
to minimize McClernand's role in the operations was to join the army
himself, as overall commander in the field.

At this point, all Federal attempts against Vicksburg had been costly failures in both the efforts wasted and the men lost. Grant decided the next attempts against the Southern citadel should be more creative.

Experiments with Sedimentary Geology

After Grant's initial failure to capture Jackson, Mississippi, and Sherman's repulse at Chickasaw Bluffs, the overall commanding officer decided to try something new—or rather, to try multiple new approaches to move his army toward Jackson and eventually Vicksburg. The primary obstacle to this maneuvering was the 100 miles of loess bluffs rising along the edge of the Mississippi floodplain. Bogs, bayous, and oxbow lakes blocked an overland approach down the valley, and transport by boat was essentially eliminated as long as the Confederates had guns high above the meander cutbanks. Confederate cavalry, familiar with the soft ground, didn't help matters as they remained a constant threat to Union supply lines.

Thus, Grant needed his engineers to find a path to central Mississippi that could be taken by boats, but not simply directly down the river. Starting in December 1862 he put the most creative of these sappers in charge of finding a watercourse; his engineers did not disappoint, providing not one but four separate proposed approach strategies. Each plan exploited the local sedimentary geology well: the spade and the temporary bridge replaced the rifle and cannon in importance for the army during early 1863. While the Army of the Potomac was being routed at Fredericksburg and Chancellorsville, the Army of the Tennessee was digging.

The four distinct and independent plans were commenced roughly concurrently.[13] Each shared one characteristic: if successful, Grant's army would be transferred from Milliken's Bend—a supply base on the west bank of the Mississippi 20 miles north of Vicksburg—rapidly and without significant loss, to the outskirts of Jackson, Mississippi.

Grant was skeptical of the first of the proposed experiments, a renewing of the bypass canal across the point bar below Vicksburg. Nevertheless, this attempt to bypass the Vicksburg batteries was especially popular with Lincoln.[14] So work was restarted on constructing a canal across the De Soto Peninsula, transforming the peninsula into an island and the meander into a lake. What Brig. Gen. Thomas Williams and three thousand men couldn't complete in July, William T. Sherman and twenty thousand men would attempt in February of the next year.

FIG. 13.4. A detail from a sketch in Frank Leslie's illustrated newspaper from March 28, 1863 of Grant's Canal. This sketch, by Henri Lovie, illustrates the sloppy nature of the canal bed—soldiers needed to use wooden planks to avoid sinking into the mud. Library of Congress.

Sherman also had several reservations about the plan. First, the upstream end of the canal was projected to connect to the river at an eddy—a small countercurrent where the whirlpool-like water flow might not allow the river to enter the canal at the velocity and volume needed to help dredge sediments from the canal bed. Second, the south end of the canal was planned to rejoin the Mississippi at a point that was within artillery range of the larger rebel guns on the bluffs of the city waterfront. Also, it hadn't stopped raining for a week.

Two dredges were brought south to speed up the digging of the canal, and both received attention from the long-range Confederate artillery pieces along the southern cliffs of the city. While the heterogeneous floodplain sediments were easy enough to excavate (without explosives), the variety of sand, silt, and clay layers provided a number of different engineering problems. Sand creates highly permeable layers, very common in point bar deposits, that increased seepage of river and groundwater into the canal. The men digging in the ditch had to work in several feet of slop (see figure 13.4).[15] Additionally, some layers of clay were tremendously cohesive, making them stubborn to dislodge.[16]

The canal, if ever finished, needed to be wide and deep enough to allow for the passage of a ship with a 60-foot beam and a 9-foot draft. Dams were constructed on either end of the canal in a rather futile attempt to keep water out of the ditch during construction. These cofferdams were dug by hand, and the mix of sediment used in their construction was never properly compacted.[17] The never-ending rain didn't help this situation either. The river continued to rise, sometimes at a rate of a foot per day.[18] On March 6 the flooded Mississippi attempted to bypass the meander prematurely, blasting through the upper dam. The lower dam held, however, so water began to quickly rise in the canal until it broke the artificial levees that ran parallel to the ditch. This breach began to flood the soldiers' already rain-saturated camps. Desperate to close this gap, soldiers tried to throw sandbags across the opening. As they shoveled sediment into the cloth bags, however, they found that the clay and silt of the mud simply leaked through the cloth bags. For weeks they couldn't get rid of the muddy sediment fast enough. Now, when they needed sediment for their bags, they found they had an unlimited supply of the wrong kind of clastic particles.

After ten days and repeated repairs this levee was finally, if temporarily, sealed. The canal was a muddy mess.[19] The two dredges began their work again, moving from north to south, doing, according to Grant, "the work of thousands of men."[20] As they did so the Confederate artillery fire became more effective, reducing their progress. Grant surveyed the situation, and after considering the enemy firepower available above the southern end of the canal, he called an inglorious halt to the entire project.[21] He would concentrate his attention on one of the other three experiments currently in progress.

The second of these engineering strategies also involved a canal, but this time the artificial waterway was not intended to create an oxbow lake but connect to one. Lake Providence was located 40 miles north of Vicksburg, sitting one mile inland from the current course of the river. If a canal could be dug to reach this lake from the Mississippi, ships might be able to find a path south to join the Red River, which eventually dumps back into the Mississippi more than 100 miles south of Vicksburg. A bunch of bayous (Baxter and Macon) would also need to be cleared of trees and other obstacles if this path was to be practicable. With the opening of this waterway, Grant could steam his gunboats down the bayous and the Red River, bypassing the guns of Vicksburg, to join Farragut for an attack on Grand Gulf. A successful strike here would open the river to shipping again, with only a slight detour to avoid the most formidable Confederate stronghold.

Grant placed Gen. James McPherson, an engineer by training, in charge of the project.[22] The 100-foot-wide canal from the Mississippi to Lake Providence was cut easily through the floodplain sediment.[23] Creating a water course between the bayous presented a much larger challenge. The cypress trees, not the sediment, were the primary problem. These trees and their stumps were exceedingly difficult to cut and remove, especially in standing water. Constant fluctuations in the stage of the Mississippi (and lakes and bayous) made the situation even more difficult. An ironclad might enter a bayou only to be trapped there a day later by falling water. In the end, the cypress and unpredictable river stage preordained that this bypass of Vicksburg would represent an inconsistent passageway at best, and the project was abandoned. Nevertheless, Harrelson et al. (2016) summarize the mixed results of this experiment well: "Later, this canal was referred to as the 'Lake Providence Boondoggle', but it was the only one of Grant's navigation canal projects that successfully bypassed the Vicksburg defenses."

Experiment number three began 150 miles north of Vicksburg. In the 1850s the state of Mississippi constructed a massive levee—100 feet thick and 18 feet high—across the Yazoo Pass. This levee ended shipping along the pass and into the Tallahatchie River, but this waterborne trade was not deemed as important as preventing flooding from the Mississippi's water into an agriculturally important valley downstream. Grant's engineers proposed to blast open this levee so that the army's ships might transport troops through Moon Lake (an oxbow lake) and pass into the Tallahatchie River, from whence they could steam toward Vicksburg from the northeast. On February 2, just as the flooding river was giving the canal diggers to the south all manner of problems on the De Soto Peninsula, the Federal engineers set off their explosives at the base of the levee. The river's water level was 9 feet above the pass at this flood stage, and the blast and floodwaters immediately cut a 200-foot-wide gap through the fractured earthen embankment.[24]

Five days later the water velocity was still dangerously fast, but the discharge had slowed enough that it was determined that barges might be able to survive running the rapids through the former levee and pass from the Mississippi into the lake. Thus, forty-five hundred men were piled onto large flat-bottomed rafts and set loose to careen from river to oxbow.[25]

Once successfully in the lake, the barges and accompanying paddle wheelers began to move south. The Confederates, anticipating this maneuver, cut trees at every opportunity to fall across the path of the ships. They also hurriedly constructed an earthen fortification named Fort Pemberton at the intersection of the Tallahatchie and Yalobusha Rivers.[26] This small fort was well placed and planned—it was built along a narrow pass of the river where only one or two boats at a time could approach, and it was surrounded by marshy, impassable floodplain terrain.[27] It was also hidden behind a meander, providing the fort with an ambush position and giving the cannoneers an element of surprise over approaching enemy vessels.[28] The well-considered position of the batteries also acted to minimize the range of any gun duel with approaching vessels. Finally, it was fortified by parapets of cotton bales and sandbags, the latter of which were better at stopping a bullet.

This fortification proved its value immediately on the arrival of the first two Federal gunboats. When their attempt to steam past the batteries failed, the entire expedition was called off. If armed vessels couldn't get by the fort, infantry-laden transports had no chance of survival. Federal infantry also had no way to attack Fort Pemberton across the muddy, saturated bayou terrain, and the ships had no way to run past the fort's guns. The fort had provided the deterrent necessary to turn the barges and paddle wheelers around for a return trip to Moon Lake.

Grant's final experiment resembled the plan at Yazoo Pass, but with less blasting and more tree-cutting. This time his engineers discovered a way to access the Yazoo River via Steele's Bayou, Black Bayou, and Deer Creek. Once these shallow-water, tree-laden water bodies were navigated, the flotilla could steam along the Yazoo River and disembark troops between Yazoo City and Haynes' Bluff. This would bring Sherman's infantry into position north of the right flank of the Vicksburg defenses.

Once again the trees of the bayou, whether felled by the Confederates or not, proved a major obstacle. Acting rear admiral David Dixon Porter's boats were averaging a mere half mile per hour.[29]

The Confederates had prepared a trap for Porter's fleet in the Yazoo River. Obstacles had been constructed across a narrow part of the channel, and sharpshooters were positioned along the banks. They also began to cut down trees behind the path of the Union vessels, making escape even more difficult.[30] Porter's boats piled up on these barriers while the sharpshooters sent minié balls flying across the water. Sherman, hearing of the situation, immediately disembarked infantry and sent the soldiers across the highest ground in the bayou to dispatch the rebel riflemen and help free the boats from the obstacles. With this close call averted, Porter's fleet returned to the relative safety of the deep Mississippi River, and the Steele's Bayou experiment was left unfinished.

Frustrated by the failure of his engineers to dig and blast a bypass south, Grant now faced the most important decision of his military career: Sherman pushed him to return his army to Memphis, but his own gambling instinct was to march south along the west bank of the Mississippi. Federal morale also certainly played a part in his decision, as terrific defeats in the east and soul-crushing decisions by other army commanders (e.g., Burnside's "Mud March") made retreat an unfavorable choice for his own men.[31]

Grant discussed his new plan with Porter. He intended to march quickly to New Carthage on the west bank of the river below Vicksburg, but he needed transport ships and protection to get across the river and into central Mississippi. Could Porter steam his gunboats and transports under Vicksburg's batteries to provide a means of crossing the broad river? The rear admiral assured the general that he could, but with one large caveat: this strategic move might be more of a risk than Grant realized. The Mississippi current would aid Porter's gunboats tremendously when running downriver past the Confederate batteries, but once there, it would be exceedingly costly for the slow ships to fight both the currents and the rebel cannon to return upriver again. And Grant would now be alone in enemy territory with no hopes of supplies or reinforcements, or even artillery support from the gunboats.

On March 29 Grant rolled the dice, sending McClernand's corps south. On the night of April 16 Porter started downriver with eight ironclad gunboats and three steamers. The gunboats sailed closer to the Vicksburg batteries while the transports hugged the Louisiana shoreline.[32] The rebel artillery fire was heavy. The next morning Porter found that all of the ships had been hit by enemy gunfire, but only one transport, the *Henry Clay*, and one coal barge had been lost.[33]

While Sherman distracted the Confederate commanding officer in Vicksburg, John Pemberton, with a series of demonstrations to the north, Grant searched for a place to cross the river. He was plagued by poor maps, and his options were limited by the Confederate river batteries at Grand Gulf. The march of his infantry across the Mississippi floodplain was also convoluted: Grant had to follow a circuitous route around the outside of multiple oxbow lakes, creating a meandering line of march that mimicked the former path of the river. For much of the journey his men stayed above the flooded bayous on the natural levees of the former path of the winding river.[34] On April 30 he selected Bruinsburg as his crossing point, partly because it was poorly guarded but primarily because the roads led directly east out of town.

Once across the river, the Army of the Tennessee moved quickly, untethered from supply lines and living off the land. Grant's men won a victory at Port Gibson on May 1, taking full advantage of their superior numbers. Grant lost 850 of his 23,000 men engaged, while the Confederates had 400 men killed and wounded and another nearly 400 captured, out of their force of 8,000. This victory sealed the fate of Grand Gulf as well, and the city was abandoned as a result of the battle.[35] Grant had now established a strong foothold on the eastern side of the river. Striking north against Vicksburg immediately presented an inviting option, but this maneuver might allow Pemberton and his garrison an avenue of retreat to the northeast, so Grant decided instead to move east toward Jackson, leaving the loess and fighting on Coastal Plain clays, sands, and sandstones.[36]

After a small battle at Raymond, Mississippi, the Army of the Tennessee moved into Jackson, capturing the city after a brief skirmish. Once the rail lines that ran out of the city were destroyed, Sherman, McClernand, and McPherson all began moving their corps west toward Vicksburg. Pemberton offered resistance en route, first at Champion Hill, where he tangled with McPherson, and then briefly against McClernand along the Big Black River. The rout at this bridgehead forced Pemberton to fall back into the defensive ring of fortifications outside of Vicksburg, where he would remain for the next fifty days.

The Siege of Vicksburg: Loss on the Loess

On May 18, 1863, the Army of the Tennessee surrounded the eastern defenses of Vicksburg. The Confederate line ran for eight miles around the city, with both flanks protected by the Mississippi River. Many of the Confederate earthworks had been constructed in mid-1862, and by the late spring of the next year, they showed signs of neglect. Unfortunately for Pemberton's men, shovels and picks were in short supply.[37] Fortunately for the rebels, they had slaves available to do much of the digging and repair.

Confederate forts and artillery positions were spaced on commanding ground along the line, and wherever the terrain would allow, they were distributed so as to provide supporting fire for adjacent strongholds. The earthen parapets were 20 to 25 feet thick and constructed primarily of silty soil and disarticulated loess.[38] Between these forts, redoubts, and redans ran trenches and rifle pits, cut easily into the soft "bedrock." In total, the Confederates had 102 (field) artillery pieces scattered along seventy-seven different positions.[39] Their largest cannon, including thirty-seven of the sixty-nine available siege guns, were positioned overlooking the river.[40] The largest of the fortifications were positioned to protect the six roads running into the city.

The loess under Vicksburg weathers in such a manner to produce deep ravines between otherwise fairly flat topography.[41] These water-eroded gullies and depressions widen away from the city, and as a result, the Union siege lines enveloping the Confederate earthworks were more discontinuous: the wider ravines produced large gaps or breaks in the Federal trench lines. This topographical feature also explains why the Union army concentrated their artillery more than the Confederates—there were simply fewer advantageous and elevated positions available for the Federal guns (see figure 3.11).[42]
Grant admired the Confederates' preparation of this superlative defensive terrain:

> The ground around Vicksburg is admirable for defense. On the north it is about two hundred feet above the Mississippi River at the highest point and very much cut up by the washing rains; the ravines were grown up with cane and underbrush, while the sides and tops were covered with a dense forest. Farther south the ground flattens out somewhat, and was in cultivation. But here, too, it was cut up by ravines and small streams. The enemy's line of defense followed the crest of a ridge from the river north of the city eastward, then southerly around to the Jackson road, full three miles back of the city; thence in a southwesterly direction to the river. Deep ravines of the description given lay in front of these defenses. As there is a succession of gullies, cut out by rains along the sides of the ridge, the line was necessarily very irregular. To follow each of these spurs with intrenchments, so as to command the slopes on either side, would have lengthened their line very much. Generally, therefore, or in many places, their line would run from near the head of one gully nearly straight to the head of another, and an outer work triangular in shape, generally open in the rear, was thrown up on the point; with a few men in this outer work they commanded the approaches to the main line completely.[43]

The Union lines roughly paralleled the Confederate crescent, running discontinuously for approximately 12 miles. As the siege dragged on,

Grant would eventually mount 250 guns along his lines.[44] These included six siege pieces (32-pounders) and a battery of heavy naval guns manned by Admiral Porter's sailors.[45]

Loess is an interesting sediment from a military engineering perspective. When undisturbed the weakly cemented silt is easy to excavate through, yet it doesn't need a significant degree of revetting or bracing. Civilian and military "miners" found this to be the case when excavating underground dwellings and tunnels.[46] The relatively impermeable nature of the loess provided an underground cavern that had a relatively low humidity, comfortable for a home and good for storing gunpowder.[47] An artillery position could be cut into the reverse slope of a hill for protection, and little additional lumber was needed before the guns could be brought forward and positioned.[48] However, if the silt became disarticulated, that is, the silt grains were separated and the cement holding the grains weakly together was broken, the sediment lost nearly all of its strength. It is no wonder, then, that after the Confederates dug out and transported the loess soil to construct ditches and parapets, maintenance of the earthworks became an issue of concern within a year. The silt would weather quickly, having little clay for cohesion, and be eroded rapidly during heavy rainfall and runoff. The silty loess was also poorer at stopping a minié ball than was sand.[49]

On May 19 Grant remained ignorant of the strength of Pemberton's forces behind the Vicksburg earthworks. He assumed the rebel force, regardless of its size, was of low morale after the defeats at Champion Hill and Big Black River. To the contrary, Pemberton had two fresh divisions awaiting the Army of the Tennessee's eventual attack.

Sherman was the first to attack, against the northern sector of the Confederate defenses along the aptly named Graveyard Road. McPherson and McClernand supported Sherman's men on the left, but the entire assault was a bloody mess that did nothing other than educate Grant about his underestimation of the Confederates' strength.

Grant reconsidered his overall strategy at this point, bringing Porter's gunboats and his own artillery into the fight in a more effective manner. On May 22 these guns commenced a bombardment of the Confederate lines, and late in the morning Grant renewed the attack with his three corps. Only McClernand had any success at the Railroad Redoubt, but his men were eventually thrown back by Pemberton's fresh reserve units (see figure 13.5). In the assault Grant had lost

FIG. 13.5. This portion of the loess battleground is located between the Railroad Redoubt and Fort Garrott. The photograph is taken from the Union line across the valley towards the Rebel earthworks. The Federals constructed multiple approach-trenches across these gullies from whence sharpshooters produced many Confederate casualties along their front.

3,200 men (> 85 percent of all casualties), while only slightly diminishing the Confederate defenders' fortitude.

These two costly assaults convinced the Federal commander that a siege was now required to defeat the enemy garrison. Porter, and the broad river, would restrict any potential source of supplies from the west, and Grant would tighten his lines to the east. More digging commenced in the soft loess. In June multiple sets of saps were excavated toward the Confederate lines, always in a perpendicular zig-zag fashion to prohibit rebel enfilade fire. Multiple mines were also dug toward key Confederate strongpoints.

When one of the longest of these trenches, Logan's Approach, came within 25 yards of the Confederate line, the digging went subterranean. The Federal miners dug quickly through the loess and needed very little bracing.[50] A gallery 45 feet long was excavated along with two smaller galleries, each extending away from the main gallery at 45° and extending the space for explosives by an additional 15 feet.[51] A total of 2,200 pounds of black powder was placed in these chambers, and loess and sandbags were used to close the mine and help direct the effects of the blast vertically, concentrating the explosion directly into the underside of the Third Louisiana Redan.

During the mid-afternoon on June 3, the explosives in the mine were ignited, producing a crater 30 feet in diameter and 15 feet deep. Federal infantry charged into this chasm to find the enemy was gone. The Confederates in the area had pulled back to a new, deeper line during the excavation of the mine, and now rebel small arms fire and artillery rained down onto the men crossing the pulverized loess in the crater. Grant noted in his personal memoirs that the Confederates made great use of hand grenades during this fighting: the rebels simply rolled the small bombs down the face of the parapet, while Union soldiers had difficulty throwing the grenades back uphill out of the crater and over the earthworks.[52] Realizing the hopelessness of the situation, the Union infantry now retreated beyond cannon range to await instructions. The project wasn't completely abandoned, however, as another mine was exploded at Logan's Approach a week later.

Union miners were not alone in digging in the loess, and their subsurface activity was hardly a secret. The Confederates were busy building countermines to stop this tactic and set off at least one explosion of their own.[53]

By July 4 the Confederate garrison and the citizens of Vicksburg had had enough. Running low on food, drinking contaminated water, and left with no hope of relief, the garrison surrendered.[54] Grant had finalized a great victory the same day Lee began to retreat, thoroughly defeated, from Gettysburg. In doing so, Grant had captured an entire Confederate army, while in the east Meade would allow Lee to escape across the Potomac into Virginia.

Grant would take the sedimentary geology lessons learned in the loess with him when he moved to the Eastern Theater. At Petersburg he would dig seemingly unlimited stretches of trenches through the Piedmont and Coastal Plain soil and explode the most famous mine of the Civil War.[55] Union engineers would also return to digging canals in hopes of bypassing Confederate batteries along the James River. In many ways Vicksburg was a trial run for the engineering projects of the Overland Campaign, and one with sediment that was ideal for such tactics.

PART V

To Take the Coasts

Against the destructive effects of projectiles, pure quartz sand, judiciously disposed, comports itself unlike any other substance.

—Union general Q. A. Gillmore reporting on the sand from Morris Island

Colonel, they will make it pretty warm for you here with shells, but they cannot breach your walls at that distance.

—Robert E. Lee, underestimating the effectiveness of new rifled artillery when conversing with the commander of Fort Pulaski

CHAPTER 14

Protecting the Shoreline

Permanent versus Temporary Fortifications

Fortifications during the Civil War, and earlier nineteenth-century wars, were divided into two broad categories defined by their ease of construction and intended duration of use.

Permanent fortifications were elaborate works that were carefully planned and prudently constructed; they were intended to be occupied and provide protection for decades. The labor and time needed for such construction usually meant that they were erected during peacetime when the labor and materials required were more available and easily acquired. In the United States, the coastline and the navy were deemed the first line of defense against foreign powers, and thus the majority of permanent fortifications were built to protect major harbors, naval yards, and ports.[1] The expensive and expansive forts around Charleston (Forts Sumter and Moultrie), Savannah (Forts Pulaski and Jackson), Mobile (Forts Gaines and Morgan), Pensacola (Forts Barrancas, Pickens, and McRee), New Orleans (Forts St. Philip and Jackson), and the Florida Keys (Forts Taylor and Jefferson) were all permanent fortifications that were constructed in the decades prior to the outbreak of the Civil War for exactly this purpose.

Temporary fortifications were described by Dennis Hart Mahan, the famous military engineer and author, as such: "When the position is to be occupied only for a short period, or during an operation or campaign, perishable materials, as earth and wood, are most commonly used."[2] These fortifications are constructed under the stress of war, usually in a rapid manner, and they lose nearly all their strategic value when the fighting shifts elsewhere or the war is over. Temporary fieldworks include breastworks (piles of rocks, fencing, timber above ground), entrenchments (requiring digging), and artillery parapets (walls of soil and sediment with or without a ditch).

When properly utilized, sediments give a temporary fortification all the strength (and more) of a permanent fortification while at the same time possessing the beneficial qualities of temporary field fortifications: ease of construction and repair. As the Civil War progressed, the increased use of rifled artillery magnified the defensive effectiveness of sediments, and many permanent brick forts constructed in the previous decades began to be reinforced with exterior parapets of sand and earth.[3]

Permanent Fortifications and the Growing Obsolescence of Brick

The first permanent fortification fabricated along America's shoreline was constructed by the Spanish in 1672 to protect Saint Augustine, Florida (see figure 14.1). This imposing and impressive structure, named Castillo de San Marcos, was constructed out of coquina, a sedimentary rock composed of lime and shells. Coquina is an especially soft sedimentary rock that was thought to have the ability to absorb punishment from solid shot without quickly crumbling.[4] During the Civil War Castillo de San Marcos was captured without bloodshed by Confederate and later Union forces.

The majority of the renowned brick and stone forts from the Civil War were designed and constructed after the country's vulnerability to invasion was clearly demonstrated by the British during the War of 1812. In the peaceful decade following this war, the United States selected forty coastal sites to receive prioritized protection. Military strategists decided that America's first line of defense would be with its navy, and not necessarily complete fortification of the entirety of the country's thousands of miles of shoreline.[5] As a result, the areas to receive the strongest defenses would be the ports and shipyards.[6]

Engineers and planners were faced with multiple geological challenges when tasked with building massive brick or stone structures along the coast. A permanent structure needed a building site that would be free from the threat of coastal erosion for many decades.[7] Ad-

FIG. 14.1. Castillo de San Marcos, the first permanent fortification built along America's Atlantic shoreline. The grey coquina blocks were soft enough to absorb spherical solid shot without substantial amounts of damage, at least for a short period of time.

ditionally, finding a suitable hard building material could be a challenge on the Coastal Plain, where the closest hard rock available for quarrying was often more than 50 miles away. The closest durable hard rock to Charleston, South Carolina, for example, is more than 100 miles distant.

While igneous and metamorphic rocks are stronger and more durable, brick offers the advantage of being constructed on site, and the Coastal Plain has a nearly limitless supply of the materials needed to make bricks: sand and clay. Most masonry material also offered one more important advantage for use in construction along the shoreline: unlike timber or iron, bricks are resistant to deterioration from salt water.

At Fort Sumter, in Charleston Harbor, the use of sediments and sedimentary rocks has an interesting history. Local sediments were used to form the bricks of the fort, of course, and granite boulders brought from New England were used to provide a foundation. Inside the fort, sandstones were found throughout. Most of these red and brown rocks are almost certainly Triassic in age, being quarried in the Triassic rift basins from the Piedmont (see figure 14.2).

FIG. 14.2. This is a bricked-in waterline gun embrasure from the right face of Fort Sumter. Note cross-bedding in the block of sandstone below the former-gun opening and possible fossilized burrows to the left.

The right face of the fort, which was the farthest from Morris Island and suffered the least damage from shelling, has surviving casemates that demonstrate how sandstone was used in the construction of the embrasures. The periphery of the embrasure opening was lined with sandstone, not stronger, more durable granite. A block of granite was used to provide a strong base for the embrasure, but the igneous rock was not used where incoming shellfire was anticipated to be concentrated. Instead, softer sandstone surrounds this more vulnerable location. The sandstone was probably chosen for two reasons: it was easier to carve and craft into shape, and it was softer than the granite—it might fragment into smaller, less dangerous shards if hit by incoming fire. The blocks of sandstone have also weathered to reveal some interesting sedimentary structures (figure 14.2): the slab of rock below and to the right of the bricked-in gun opening exhibits cross- and planar bedding, and the block to the left appears to show signs of burrowing.[8] The casemates of the right face were deemed obsolete by the time of the

FIG. 14.3. Fort Sumter under the Confederate flag. These casemates would soon be turned into rubble by Union artillery, but the fort remained a viable defensive position even after the brick had been turned into coarse sediment. The most interesting part of this photograph (enlargement, below) is the partial burial of gun tubes in the rubble and sand, essentially elevating guns into mortars. Library of Congress.

fort's renovation in the late 1890s, and the entire set of casemates—guns and all—was buried in sand to strengthen the scarp.

Whenever possible, fortification architects preferred to use the brick or stone to construct arches (see figure 14.3). Arches allowed for the construction of multiple tiers, increasing the number of casemates and doubling or tripling of the firepower of a fort.[9]

Two massive brick forts, both of which had their construction initiated in 1829, offer an interesting illustration of the challenges from sedimentary geology that were encountered by engineers regarding proper site selection: Fort Sumter in Charleston Harbor, and Fort Pulaski, on Cockspur Island downriver from Savannah.

Fort Sumter's site was selected to provide cross fire over the entrance to Charleston Harbor with Fort Moultrie, a smaller fort that was located on the northern shore of the harbor on Sullivan's Island (see figure 14.4). Sumter was to be constructed on a shallow portion of the harbor's flood delta—essentially a shallow shoal of sand carried into the harbor by incoming high tides. This shoal is adjacent to the northern

FIG. 14.4. The entrance to Charleston Harbor is dominated by the guns of Fort Moultrie, foreground, and Fort Sumter, background. When fully garrisoned during the Civil War, these two forts held almost 150 guns. The sandstone-lined embrasure pictured in FIG. 14.2 lies along the right face of Sumter's scarp.

edge of Morris Island. The combination of beach drift and the shifting of the tidal delta made the water especially shallow and energetic in this area.

This depositional environment is both dynamic and complex, requiring a massive engineering effort to stabilize the foundation for the multitiered masonry building. The shifting sands would provide a challenge on multiple fronts: the friction of the sediment might support a portion of the weight through limited compression, but with little clay present, cohesion would be minimal, and the foundation of the fort would be highly vulnerable to erosion from waves, tides, and strong bottom currents.[10]

Engineers decided that an artificial island would need to be constructed for the foundation of the fort. More than 50,000 tons of rock boulders (riprap), 10,000 tons of which were granite brought by ship from New England, were imported and dumped on the shoal (see figure 14.5).

The maximum size of particle that water can pick up and transport in suspension (competence) is directly related to water velocity: fast-moving water can pick up and move coarse sand and pebbles, while

FIG. 14.5. Detail of George Barnard's 1865 photograph taken on the granite riprap foundation outside of Fort Sumter. Partially visible on the top left of the photograph is the horizontal fraise constructed by the Confederates to prevent night intrusion. Library of Congress.

slow-moving water can only carry clay in suspension. Competence is the reason riprap boulders were chosen instead of some other smaller grain size. Riprap consisting of angular igneous boulders offers a great deal of friction to support the fort's foundation, and it is nearly impossible to erode because the water velocity necessary to raise the competence to transport the boulders would be greater than those witnessed in even a powerful hurricane.

Anaconda in the East

In April 1861 Abraham Lincoln decided to use the Union navy to blockade the ports of the South, thus cutting off the economic and military benefits that international commerce might offer to the Confederacy. Two months later the Northern government established a special planning committee, the Navy Board, to determine how best to accomplish the president's blockade. One of the initial decisions outlined in the plan to arrest Southern trade with Europe was a division of the responsibilities for the navy: two blockade squadrons, one on the Atlantic coast and the other in the Gulf of Mexico, would be created to guard the ports from Alexandria, Virginia, to the border with Mexico. Key West, Florida, would mark the dividing line between the squadrons.

For the next four years the Navy Board would provide guidance concerning joint army-navy operations along the lengthy Confederate coastline. The first target of the navy would be in the Carolinas. The navy would take full advantage of the new tactical possibilities created by the introduction of steam engines. By providing their own means of power, instead of relying on the whims of the wind, the Federal ships could now exert power wherever and whenever they chose. Naval tactics were no longer reliant on favorable weather. This flexibility of maneuver was first demonstrated at Hatteras Inlet. The capture of this strategically important passageway would provide access to Pamlico Sound, quiet water that offered a degree of protection for the blockading fleet.[11]

Hatteras Inlet was protected by two sand fortification, however. The larger fort was elevated and offered firing positions across the inlet and into the sound, but it was poorly armed. The smaller Fort Clark was located across the inlet and was even more vulnerable to the seven warships and 158 guns brought south by the North.

The Confederates soon learned just how perilous the garrison in Fort Clark was: the larger, longer-ranging Union artillery allowed the Federal ships to bombard the fortification from beyond the maximum range of the Southern cannon. Within hours the entire garrison fled from the fort for shelter in the larger Fort Hatteras. The next morning the Federal ships began bombarding that sand fortification as well, while always continuing to maneuver and turn in unpredictable directions. By noon the outgunned 670-man garrison surrendered the fort.[12] The fall of the weak Fort Hatteras and weaker Fort Clark provided misleading lessons about the defensive capabilities of forts made of sand. The capture of the inlet appeared to be a demonstration of the effectiveness of the combination of the new steam propulsion and larger ordnance over sand forts—a circumstance that shaped misguided conclusions for future deadly coastal operations.[13]

The next army-navy coordinated operation was conducted by the Gulf Coast Blockade Squadron, and it proved to be equally successful with the capture of Ship Island, Mississippi. Ship Island protects the only deepwater harbor between Mobile Bay and the Mississippi River, rendering the island an inviting target.[14] The small barrier island had been used by Maj. Gen. Edward Pakenham as a base of operations for his fleet of more than fifty ships as he prepared to attack New Orleans in 1815.[15] The island was vulnerable to occupation, protected only by a single unfinished brick fortification: Fort Twigg. Soon after the Federal screw steamer USS *Massachusetts* bombarded the fort on July 9, 1861, the island was abandoned. The diminutive fort was renamed after the ship that had captured it, but it still wasn't completed when the war ended four years later.

With the capture of Ship Island, the Gulf Coast Blockade Squadron gained two valuable resources, as described in the January 4, 1862, edition of *Harper's Weekly*: "Excellent water can be obtained in unlimited supply by sinking a barrel anywhere on the place. The great advantage of this is too palpable to require comment. The island possesses a very superior harbor. . . . The rise and fall of the tide is only from twelve to fourteen inches."

So, by the end of the summer of 1861 the Union navy had accomplished the first two tasks recommended by the Navy Board: capture Hatteras Inlet and take Ship Island. The seizure of Hatteras Inlet greatly restricted Confederate commerce in the sound, but it did not provide a satisfying deepwater anchorage for the navy. Next, it was decided, the navy needed to acquire a better port on the Atlantic for its deeper-draft vessels. The earlier coastal success had been primarily, or entirely, accomplished by the navy; the army had taken a secondary role behind the heavy artillery of the ships. This next operation was intended to be the first truly joint operation between the land and sea forces.

The ideal target for such an operation would be equidistant between Norfolk, Virginia, and Key West, Florida. Charleston, South Carolina, was protected by the intimidating Forts Sumter and Moultrie. Savannah was guarded by Fort Pulaski and, to a lesser extent, Fort Jackson. Nearby Port Royal Sound and Beaufort, South Carolina, however, resembled Hatteras Inlet with respect to the level of fortification and protection: the sound was only guarded by two smaller sand fortifications. Fort Walker was located on Hilton Head Island and held sixteen guns. Fort Beauregard, sited on Bay Point, had eight smaller cannon.

Opposing these two forts would be the South Atlantic Blockade Squadron, commanded by Samuel F. DuPont and thirteen thousand troops led by Brig. Gen. Thomas Sherman.[16] The flotilla put to sea at the end of October 1861, and the voyage south greatly reduced its formidability. Storms wrought havoc on the ships, sinking much-needed supplies of ammunition and destroying landing craft. By the time the strike force was assembled on November 5, "the other" General Sherman informed DuPont that his infantry were unavailable.[17] Without a viable way to get the foot soldiers ashore, there would be no "joint" in joint operations.

DuPont determined to go it alone, employing the tactics used at Hatteras Inlet. His ships would be constantly in motion and would engage both forts when entering the sound while circling counterclockwise, continually firing.[18] After a reconnaissance raid to map the depths of the sound and determine the navigability of the planned attack route, the main battle force moved forward on the morning of November 7. The battle was short-lived. Fort Walker was the first to surrender; her guns had occupied an elevated position on the sand parapet to give them increased range, but this also made them more vulnerable to close- and medium-range shellfire from the water.[19]

Quincy Gillmore, an army officer with a military specialization in shoreline engineering, pointed out that the main weakness of the forts was their design, not their construction material. The Confederate garrisons fled due to the volume of Union gunfire, before "the works themselves had sustained any material damage." Gillmore also pointed out that neither small fort contained a bombproof, a feature that would later make Battery Wagner at Charleston nearly impervious to his own heavy artillery.[20]

With the abandonment of Fort Walker, the command of the other fort began to fear complete annihilation. Phillips Island, the site of Fort Beauregard, is a peninsula, unlike Hilton Head Island. Once Fort Walker fell, the Federal gunboats had the option of sailing into the Beaufort River, in a position landward of Fort Beauregard, thus trapping the remaining garrison in the sand fort. Predicting precisely such a maneuver, Col. Robert Gill Mills Dunovant ordered his men to retire to the mainland, abandoning the final fortification that protected the valuable harbor.

In a foreshadowing of the struggles for the sand fortifications outside Savannah (Fort McAllister) and Charleston (Battery Wagner), the ship-versus-fort engagement caused surprisingly few casualties. Only eleven men were killed in the two forts, with another fifty or so wounded. On the ships only eight Yankees were killed and twenty-three wounded.

With Walker and Beauregard captured, the Confederates left Hilton Head, and Port Royal Sound was converted by the Union navy into a coaling and provisioning station of the highest order. The Federals had quickly obtained an ideal port at little cost, located between their next two targets of Savannah and Charleston. They had also acquired a harbor large enough for the entire Union fleet in the Atlantic. The Port Royal base was further enhanced from a geological perspective: the Federal land forces on the barrier island were protected from the mainland by a vast array of impassable salt marshes, which offered insurance against a rapid surprise Confederate advance.

With this protected harbor now established, the Union navy and army began to consider a potential strike against a more important Confederate port. By mid-1862 New Orleans, the largest city in the South, was captured. When Flag Officer David Farragut first bombarded then fought past Fort Jackson and Fort St. Philip on the Mississippi River, the city was doomed. The levees that "protected" the city from flooding also raised the river's elevation; higher floodwaters elevated the firing position of the Federal gunboats as well, making the city even more vulnerable to bombardment. When New Orleans capitulated, the vulnerability of other, smaller cities was now apparent. In a direct response to the surrender of New Orleans, Pensacola, Florida, was abandoned. Nevertheless, it would take another two years for Farragut to damn the torpedoes and close Mobile Bay, the last significant port along the Gulf coast east of the Mississippi.

Along the Atlantic, the ports and prizes would be costlier. Charleston and Wilmington were the largest cities in their respective states in 1861 and, collectively with Savannah, were undeniably important ports. They were also the most heavily fortified cities along the Atlantic seaboard (see figure 14.6).[21] The Union blockade had diminished, but not eliminated, incoming supplies and weapons, and Lincoln wanted both ports closed completely. The capture of Charleston also had a second, symbolic importance in addition to its strategic value. The Union leadership and rank and file collectively viewed South Carolina, and Charleston specifically, as the provocateur for the war itself and deeply desired vengeance.

Retribution could wait, however, and Savannah was the first port to be targeted; Charleston and her forts would be dealt with during a later campaign. It would take until January 1865 for the final target of Wilmington to be closed during the concluding phases of the war. For the Confederacy, Charleston Harbor marked the birthplace of secession. Four years later the fall of the Cape Fear River would, for all practical purposes, mark the cessation of the secession.

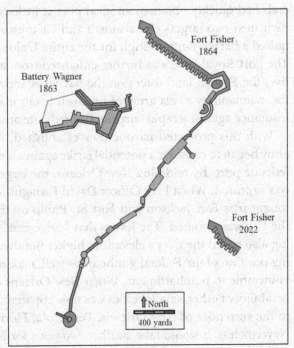

FIG. 14.6. Battery Wagner and Fort Fisher were the most troublesome sand forts defending Charleston and Wilmington for the Federal Army and Navy. At a comparable scale, Fort Fisher's imposing size gives good reason to its nickname as the "Gibraltar of the South. " Only a small fragment of this fort remains today, and Battery Wagner has been completely eroded into the Atlantic.

Fort Fisher
1864

Battery Wagner
1863

Fort Fisher
2022

North
400 yards

Shore Batteries versus Ships

The Union army and navy learned early in the war that siege artillery and naval guns were more effective when positioned on land, compared with those on naval vessels. Land batteries offered a stable firing platform, increasing accuracy over heavy artillery sitting on the rolling gun decks of a ship. Even a miniscule rate of roll could throw the range and accuracy of a gun off by hundreds of yards. Additionally, the size of a piece in a land battery was essentially unlimited, as was the supply of ammunition in a well-supported position.[22]

When faced with the enormous firepower available in Confederate-held coastal fortifications, the navy adopted two strategies. First, if the fortifications had been constructed to protect a strategically important target—a city, naval base, or harbor entrance—the Federal ships could run the gauntlet of the fortification's guns, while putting down suppressive fire, to bypass the forts and menace the strategic target. The second option required gathering an overwhelming amount of naval-borne firepower to reduce the fighting capacity of the brick or sand for-

TABLE 14.1. Comparison of the available firepower in ship versus shore engagements, listed in chronological order

The Federal navy was vastly outgunned when attacking Fort Sumter with ironclads, while during the bombardment of Fort Fisher they amassed nearly six hundred guns on seventy ships.

Engagement	Date	Ratio of Confederate guns to Union guns
Battle of Fort Hatteras/Clark	August 1861	1 : 10
Battle of Port Royal	November 1861	1 : 1.5
Battle of Fort Jackson/St. Philip	April 1862	1 : 1
Battle of Charleston Harbor	April 1863	12 : 1
Battle of Plymouth	April 1864	1 : 2.5
Battle of Mobile Bay	August 1864	1 : 2
Battle of Fort Fisher	December 1864	1 : 3

tifications. Both approaches are apparent when comparing the ratio of available naval and fortress artillery firepower during multiple coastal engagements (see table 14.1).

Ships were clearly more vulnerable when receiving artillery fire. Land batteries could disable a ship by penetrating the hull of a ship; damaging the sails, rigging, or rudder; or wounding and killing the crew.[23] For a ship to reduce the fighting capability of a land fortification, conversely, a ship would either need to reduce each artillery position or compromise the fort's magazine. With a requirement for such precision firing by shipborne smoothbore weapons of the time, this would require that the range to target be reduced to under 1,000 yards, making the ships especially vulnerable to defensive artillery fire, especially of the plunging variety.

Quincy Gillmore understood well the vulnerability of ships to gunfire arising from the land, and his approach to reducing the Confederates' coastal fortifications would eliminate these disadvantages by moving his big guns to terra firma. He had learned some important lessons about attacking permanent fortifications at Savannah and temporary shoreline fortifications at Hilton Head, but his bitterest lessons were yet to come a bit farther up the shoreline.

CHAPTER 15

The Education of Quincy Gillmore

Gillmore the Coastal Engineer

Quincy Adams Gillmore was born in 1825 and named for the then president-elect, John Quincy Adams. He was trained as a coastal engineer, graduating first in his class at West Point in 1849. At the outbreak of the war his background led to an appointment on "the other" Sherman's staff, and he participated as the chief engineer during the successful operations around Port Royal Sound. After promotion to brigadier general, he was tasked with developing a plan to capture Fort Pulaski, the brick citadel on Cockspur Island that protected the mouth of the Savannah River.

In 1829 a young military engineer named Robert E. Lee was sent to Cockspur Island along the Georgian coast to survey the region and determine a feasible site for construction of a multitiered brick fort.[1] Lee faced a problematic choice, balancing the range of the fort's artillery to the Savannah River with the long-term threats posed by coastal storms and erosion. The selected location needed to be close enough to the river that Pulaski's guns could sink any ship headed upstream toward Savannah but distant enough from the beach and river that coastal erosion would not be a threat during the planned effective lifetime of the fort.

For nearly a decade construction proceeded, often hindered by hurricanes and tropical storms. The eventual location for Fort Pulaski contrasted with that of Fort Sumter because the latter was built on sand on a shoal. Instead, the selected construction site downriver from Savannah was in the middle of muddy Cockspur Island, in the center of an expansive marsh. To stabilize the massive fort, wood pilings were driven through the compactable mud and into the coarse underlying Pleistocene sand.[2] Even so, the compaction of the mud led to settling of some portion of the massive fortification, and the proposed third tier to the works was never added to the imposing fort.[3]

Nevertheless, a 25-million-brick fort could not be supported by the mud and silt on the surface of the island. Cockspur Island was a remnant of the Savannah River delta, and the island's sediments contained more silt and mud than the surrounding barrier islands. On Tybee Island, to the south, the sand and mud were distinctly separated by subenvironment: marshes were muddy and beaches and dunes were sandy. On Cockspur, the marshes were still muddy, but the composition of the "mainland" at the center of the island was far from clean sand.[4] Instead, the best terrain on the island was a mix of mud and delta sand.[5]

Differential settling and vertical and horizontal expansion and subsidence would pull the bricks apart even before construction of the 7- to 11-foot-thick walls could be finished. To support this weight, 70-foot-long pilings were driven into the mud to provide support for a timber subflooring for the brick. These pilings penetrated through the Satilla Formation, a layer of sediment that is less than five thousand years old, to reach the Tybee Phosphate Member of the Coosawhatche Formation. This denser, tougher, older sediment would better support the masonry formation.[6]

Despite these elaborate supports for the foundation of the fort, Pulaski would never gain multiple tiers like her cousin to the north in Charleston Harbor. Pilings driven into mud and muddy sand would fail to equate to the support provided by the sand and granite riprap laid down for Fort Sumter.

The construction of Fort Sumter and Pulaski demonstrated that formidable brick fortifications could be built on unconsolidated sediments, but only on sediment possessing the friction to withstand the great degree of deformation an almost 12-foot-thick brick wall would create. Thick brick walls do not bend when experiencing differential subsidence. Because sand has a high degree of intergrain friction from the interlocking rough particles and quartz is a hard mineral, it satisfies these requirements; nevertheless, sand alone lacks the cohesion of clay and thus is particularly vulnerable to erosion from wind and water currents. To withstand such erosion, sand must be stabilized, either from rock riprap (Sumter) or an overlying muddy marsh (Pulaski).

The fall of Port Royal convinced Robert E. Lee, then acting as a special advisor to Jefferson Davis, that Georgia's barrier islands were especially vulnerable to attack.[7] He ordered the temporary field fortifications on the beaches to be destroyed and any newly surplused ordnance to be moved to Pulaski. This decision made the old brick fort as vulnerable as the temporary fortifications on the sea islands; now the Union army could establish a base for long-range bombardment across the river from sandy Tybee Island.

In early 1862 Gillmore was ordered to take the two primary fortifications protecting the marine approaches to Savannah, Georgia. These fortifications included the massive and imposing brick Fort Pulaski, completed in 1847, and the newer and smaller earthen Fort McAllister located to the south of the city. The reduction and capture of these forts would serve two purposes: it would restrict the number of blockade-runners coming in and out of the Forest City, and it would allow for a critique of the tactics planned for the eventual capture of a much more important prize—the fortifications around Charleston.

Both Confederate and Union leadership vastly overestimated Fort Pulaski's strength, considering it invulnerable to long-range artillery fire from across the Savannah River.[8] Nevertheless, Gillmore was determined to attack the fort and spent nearly three months preparing his artillery positions on Tybee Island, the closest stable, solid-earth firing platform for siege artillery. While the island was not situated at an ideal (shorter) range for bombardment, all other more proximal positions were marshlands with compressible soft mud.

Quincy Gillmore went to work covertly constructing batteries on the northern edge of Tybee, a little more than a mile from the Confederate fort. In November 1861 Lee returned to Pulaski for an inspection and to confer with the fort's new commander, Charles H. Olmstead. Lee understood the potential artillery threat from Tybee Island but underestimated its magnitude, assuring Olmstead, "Colonel, they will make it pretty warm for you here with shells, but they cannot breach your walls at that distance."[9]

Lee's opinion of the fort's strength was shared by those on the other side of the impending fight: Gen. Joseph Totten, the U.S. chief of engineers, confidently remarked, "You might as well bombard the Rocky Mountains."[10]

By early April 1862, Gillmore had constructed sand batteries for thirty-six guns, ten of which were heavy James or Parrott rifles. The batteries were concentrated along the dune-field edge of Tybee, where the friction inherent to the sand would support the weight of the ordnance (some pieces measured in at more than eight tons), and areas of the muddy marsh were avoided during construction.

After demanding surrender of the fort on the morning of April 10 and being refuted, Gillmore commenced firing at Pulaski's thick brick walls. Soon after the fort's scarp began to suffer observable damage from the incoming rifled shells. Each successive strike blasted a small crater in the brick, and by early afternoon the next day, successive hits on the same portion of the wall began to crumble the ramparts.

Pulaski returned fire, but most of the counterbattery shellfire missed the Federal guns, and even when the rounds struck the parapets, they buried themselves harmlessly in the sand. Author David Page summarized the disparate display of hitting power and resulting damage succinctly: "The effectiveness of sand and earth over stone and brick could have no better demonstration."[11]

The rifled artillery was turning brick into sediment, and the fort was becoming more vulnerable to capture or a catastrophic magazine explosion after each of Gillmore's 5,275 rounds (see figures 15.1–15.3). Sand and brick debris now filled the moat, reducing yet another obstacle for a Federal over-"land" (marsh) assault.[12]

FIG. 15.1. Enhanced detail of a Timothy H. O'Sullivan photograph of Fort Pulaski soon after bombardment and surrender in the spring of 1862. Library of Congress.

FIG. 15.2. Fort Pulaski today, after extensive repair. Newer brick repair is in the same region of the fort as "the breach," imaged in Figure 15.2.

FIG. 15.3. A second view of the eastern wall of Fort Pulaski today. Note unexploded and solid bolt ordnance still embedded in the brick wall (insets).

By the late afternoon the next day, a crevasse had widened to the point that Federal rounds were striking the opposite interior walls. The most vulnerable portion of Fortress Pulaski was the powder magazine located beneath the north-wall ramparts. The Confederates were aware of this threat, having piled wood and sediment against the interior wall of the fort to diminish the effects from shellfire from the south. They also excavated huge furrows from east to west across the parade grounds to discourage ricocheting shells from the same direction. Nevertheless, a great chasm had opened in the south wall of the fort, and this breach allowed direct fire toward the four hundred kegs (40,000 pounds) of black powder in the north magazine. Fearing a direct and cataclysmic hit, Olmstead ordered the fort's surrender. At this point, the fort was exceedingly vulnerable to a catastrophic explosion from its now threatened gunpowder supply, and the surrender of the fort was accomplished without the need for a costly infantry assault.

Subsequent to the fort's fall, an important lesson was learned by commanding officers on both sides: rifled artillery effectively destroyed a masonry fortification that had previously been thought impervious within thirty-six hours, and the return fire from the fort had little, if any, appreciable impact on the sand parapets and batteries on Tybee Island. Masonry forts might be able to hold their own against naval gunfire, but they were doomed against large-caliber land-based rifled artillery sitting behind sand.

In his official reports Gillmore noted that four hundred fewer pounds of rifled artillery shells were required for breaching one linear foot of brick than when using smoothbore artillery.[13] At similar ranges, he went on to calculate, rifled shells penetrated approximately twice as deep into Pulaski's walls, all while being more accurate and having the defensive benefit of being fired at even greater ranges. Many historians have pointed to the fall of Fort Pulaski as the end of the era of "impervious" masonry fortifications.[14] On an equally important note, the bombardment also demonstrated the growing obsolescence of smoothbore siege and naval guns. What had not been made as clear was the defensive strength of the sand batteries on Tybee against rifled artillery because Fort Pulaski had only one rifled gun in operation. As the Union army and navy moved north, a new strategy emerged that would incorporate these lessons: during the subsequent campaign to take Charleston, land-based artillery would need to be established within range of the city's protective fortifications, either Fort Sumter or Fort Moultrie, and the batteries could be constructed of, and on, the sand of the islands.

Gillmore's theater commander, Maj. Gen. David Hunter, reported the shocking impact, so to speak, of the new rifled artillery to Washington, D.C.: "The result of this bombardment must cause a change in the construction of fortifications as radical as that foreshadowed in naval architecture by the conflict between the Monitor and Merrimac. No works of stone or brick can resist the impact of rifled artillery of heavy calibre."[15]

It was obvious to all present that the new rifled artillery was lethal against brick structures, and the permanent fortifications constructed during the previous decades of peace were now, for the most part, obsolete. What wasn't so obvious was the effectiveness of sand against the same type of ordnance. U.S. Army lieutenant Horace Porter described the incoming Confederate shellfire on Tybee Island as resembling "a swarm of bees," yet not a single Federal gun was silenced.[16]

Another important lesson about the value of sand versus brick in fortification construction involved the ease of battle-damage repair. A lull in fighting would allow for quick repair of a sandy rampart or embrasure by unskilled labor; the same cannot be said for forts made of stone or brick.

Gillmore's success on Cockspur Island earned him a promotion to major general but also a transfer to the Army of the Ohio, where he would command an army division in the operations throughout Kentucky.[17] A year later, while on a leave of absence, Gillmore petitioned the Federal leadership for his return to the coast. He had caught wind of an impending move by the army and navy against Charleston, and he longed to return to the shore and repeat the success he had had along the Georgia Bight.

Geologic Challenges to Moving and Siting Coastal Artillery

Union general Gillmore had little choice but to site his closest battery almost a mile away from Fort Pulaski because of the combination of geomorphology and the need for concealment of his guns. No firm or solid land existed closer to the fort adjacent to the river or the marshes. An artificial solid platform might have been constructed in the marsh—in a manner similar to the Swamp Angel battery behind Morris Island, South Carolina—but not within range of the guns in the imposing fort.

Barrier islands, including the 2.5-mile-long Tybee, are surrounded by the ocean on one side, marshes and lagoons on the other, and inlets (or, in this case, the Savannah River) at either end. These islands tend to be longer and thinner in North Carolina before assuming a "drumstick" shape in South Carolina and Georgia.[18] Of all the island's sub-environments, only the center dune/beach portion of the island has the sand supply to successfully withstand the soil compression created by the weight of heavy artillery. Gun crews also need space to operate their weapons, and on a narrow barrier island this often restricted the potential artillery positions to the dune fields, the highest elevation close to the center or rear of the island. Such positioning was true for the siege guns aligned against Fort Pulaski, Fort Macon, and Battery Wagner on the north end of Morris Island.[19]

Beauregard and the Barriers

In Charleston the commanding officer in charge of the city's defenses, P. G. T. Beauregard, heard reports of Pulaski's decomposition before the rifled guns of the Union army and began to reconsider his strategic plans. Beauregard determined the need to strengthen the city's defenses. Sand and soil would be primary building materials, with a defensive line constructed across James Island to protect the city from a Federal advance from the south. New sand forts were constructed on Morris Island, and sand was added to the brick scarp of Fort Moultrie.[20] Unfortunately, sand reinforcement would not extend to Fort Sumter: the fortification's occupation of nearly every square foot of its artificial riprap island precluded the addition of a protective sandy parapet (see figure 14.5).

The largest and most complex of Charleston's sand batteries was constructed across the width of Morris Island, the barrier island that marked the southern border of Charleston Harbor. The Neck Battery, as it was originally called, carried a descriptive name.[21] The fortification crossed the narrowest portion of Morris Island, spanning from Vincent's Creek in the back-barrier marsh to the strand line of the Atlantic Ocean. The battery design took full benefit of the natural geomorphological advantages each of the Morris Island sub-environments offered.[22]

Whether modern or ancient, a barrier island is the subaerial accumulation of sediment between the shoreface and a back-barrier lagoon or marsh, and between two inlets.[23] A littoral (nearshore) sand body must have six interactive sedimentary environments in order to be characterized as a barrier island: (1) mainland; (2) inlets; (3) back-barrier lagoon (or marsh); (4) shoreface; (5) barrier platform; and (6) inlet deltas.[24] The spatial distribution of these sedimentary environments dictated the strategy of the advancing Union forces during the assault on Charleston in 1862 and 1863, and Confederate military engineers took full advantage of the favorable island geomorphology when constructing the city's defenses. The Federal forces were largely constrained in their movement by the narrowness of modern and Quaternary barrier island complexes. The width of an island is a function of the proximity between back-barrier marshes (or lagoons) and the shoreface. The back-barrier marsh is a vegetation-rich depositional environment that separates the mainland from the barrier island and, during periods of spring or high tide, was impassable by large bodies of troops. Whether a barrier island is backed by a marsh or a lagoon is a complex function of sediment supply from both the mainland and the island (via overwash or inlet deltas), the rate at which the island is retreating, and the inherited bathymetry offshore.[25] Assaults over any terrain other than a high marsh sub-environment during low tide would have been difficult for troops in the best possible conditions, and impossible while encumbered with equipment or under hostile fire. Moreover, movement across any marsh sub-environment would have been impossible for artillery or cavalry. In nearly all cases, the marsh eliminates flanking or enveloping maneuvers from at least one direction.

Barrier inlets are shore-perpendicular channels that separate one island element from another or from a laterally adjacent mainland constituent.[26] Without this depositional environment, the barrier island would be classified as either a baymouth barrier or, more commonly, a barrier spit. The inlets along the Charleston coast can be classified as either fluvial barrier inlets or tidal barrier inlets, depending on the relative influence, strength, and interaction of the local rivers or tides. The nature of the inlet, whether tide- or river-dominated, was an important consideration for movements of both troops and shallow-draft ships along the shoreface. The ebb-tidal delta associated with larger inlets prohibited movement of anything other than very shallow-draft ships when near the islands, making naval support of land operations more challenging. The water velocity associated with larger tide-dominated inlets generally prohibited cross-inlet movement by troops, as the boats would have been carried in a shore-perpendicular direction offshore (or, in a rising tide, toward the mainland).

Nearly all the fighting around Charleston took place on a Pleistocene barrier island complex, James Island, and a modern barrier island, Morris. Another modern island, Folly, served as a staging ground for the assaults.

A number of Pleistocene barrier islands are located between Folly and Morris Island and the mainland. These smaller and older barriers and beach ridges include Cole's Island, Black Island, Long Island, and Sol Legare Island. All of these ancient barriers played a role in the Union strategy to take Charleston, and many were fortified with artillery. The sight lines across the flat marshes offered terrific fields of fire. Transgressions and regressions over the last several interglacial cycles have left a series of stair-stepped marine and estuarine terraces along the Coastal Plain in this region, and at least six barrier ridge systems have been identified behind the modern barriers.[27]

James Island is located the farthest from the modern barriers and the Atlantic Ocean and is separated from Folly and Morris Islands by an expansive salt marsh. James Island represents the oldest and least dynamic of the islands because of its remoteness from wave energy or tidal currents. This island would also have had the largest supply of fresh groundwater and timber during the Civil War, although local plantations had clear-cut many of the forests.[28]

The shore-parallel orientation of the modern and Pleistocene barrier-island complexes dictated that any direct assault on Charleston by the Union army must come from either the northeast (Sullivan's Island and Isle of Palms) or the south/southwest (Morris and Folly Islands). Advancing from the northeast had severe disadvantages. The Pleistocene barrier-island systems behind Sullivan's Island are not as extensive or as expansive as those behind Folly and Morris Islands—any large-scale troop movement would need to take place solely on the modern islands, and this would restrict tactical flexibility. Additionally, the presence of Fort Moultrie—a strong fortification with sand-covered brick walls and a heavy artillery garrison—on the southern extension of Sullivan's Island discouraged assault on the city from this direction.[29] To the south, the expansive marsh system behind Folly and Morris Islands also limited the Union's corridor for attack.[30] The combination of short cordgrass (leaving clear fields of fire), low relief (providing long-range visibility), and wet, muddy sediments created an impassable barrier for infantry, artillery, or cavalry. All of these factors challenged Maj. Gen. David Hunter as he planned his advance along the Stono River in June 1862. His would be the first attempt to take Charleston from the south along the shoreline, and a year later, many of lessons from his geomorphologically challenged failure to capture the city would go unheeded by Quincy Gillmore.

The 1862 campaign to take over Charleston began with the capture of Edisto Island, providing a foothold for the infantry's approach to the city. This assault would shift between modern, late Holocene, and late Pleistocene barrier islands during the approach. The Union navy supported these maneuvers by bringing warships into the Stono Inlet and River. The army's plan was to move from two modern islands, Edisto and Seabrook, onto Johns Island (a complicated series of Pleistocene shorelines and islands), take Folly Island (modern island) and Sol Legare (late Pleistocene), before crossing James Island (Pleistocene) and Long Island (late Pleistocene) to capture the southern end of Morris Island (modern). From Morris Island and the adjacent Pleistocene barrier ridges, the Federal land-based heavy artillery could reduce Fort Sumter, as it had Pulaski, allowing the navy to finally enter Charleston Harbor. The presence of Federal warships adjacent to the Charleston Battery would force the capitulation of the city.

One significant complication preordained failure for this island approach: P. G. T. Beauregard had ordered the construction of earthworks across the entire expanse of James Island, including placing a sand battery across the most viable path between Folly Island and Morris Island. This fortification was called the Tower Battery because of the observation platform that overlooked the expanse of flat marshes between James, Folly, and Morris Islands. The Federal assault on this sand battery, and the role of salt marshes in minimizing Union tactical options, would provide foreshadowing—on a smaller scale—of the disastrous events to come a year later and less than a mile to the south on Morris Island.

The Confederates were well prepared for a Union maneuver north from the islands around Stono Inlet, taking full advantage of the limitless supply of local, easily movable sand. Their fortified line ran from the Ashley River and Fort Pemberton across the entire width of James Island. From here, the line continued past the Tower Battery and ended at Fort Johnson, an earthwork that overlooked the harbor and Fort Sumter.[31] At the center of this line was the village of Secessionville.

The Federal attack would center on the nearby Tower Battery, which was constructed across a half-mile-long sandy ridge. Marshes covered both flanks of this ridge (and battery) to the north and south. The commanding officer in the fort, Col. Thomas Lamar, would lead the defensive resistance in the stronghold that would later adopt his name.

The Tower Battery was located at a geographically important site: it occupied a late Pleistocene ridge that spanned a little over 100 yards from southwest to northeast. This nondescript sandy rise provided one of the few solid-earth corridors across the back-barrier marshes behind Folly and Morris Islands. Any commander determined to place artillery on Morris Island would find it easiest to cross this sandy extension; otherwise bridges or boats would be needed to traverse the marsh creeks or Lighthouse Inlet. The Confederates held the battery and this key position with approximately five hundred men, most of whom hailed from South Carolina.[32] To further fortify the site, several pieces of heavy artillery were provided to the garrison.

Maj. Gen. Hunter began to move north with the divisions of Brig. Gens. Horatio Wright and Isaac Stevens, around seven thousand men. The first significant attempt to breach the Confederate entrenchments occurred at Secessionville on June 16, 1862. During this fight, the immensely outnumbered Confederates benefited from the local geomorphology at almost every scale. The distribution of Pleistocene sand ridges and impassable muddy marshes dictated the Federal route of approach and restricted the size of the attacking infantry force. As a result, the Confederate artillery was completely concentrated along the axis of the ridge, entirely in one direction, understanding that the probability of attack from the marshes to the north or south was negligible. The flat marshes and linear, gently sloping sandy ridges also provided excellent fields of fire for the Confederate guns. This small change in relief also resulted in little potential cover or concealment for the assaulting Union infantry. As the Seventh Connecticut, Eighth Michigan, and Twenty-Eighth Massachusetts began their attack, the men discovered that the Pleistocene ridge narrowed as it approached the battery, with marshes encroaching on both sides. As the assault continued down the ridge, the Federal line became more and more constricted and the lines became entangled, with the flanks dragging across the softer, muddier ground.[33] Despite these geological challenges, the Union infantry still managed to reach the sandy parapet, but with a much diminished striking force. Fierce fighting took place between a small number of Fed-

eral infantry and Confederate artillerymen on the parapet walls and superior slope, and a Union flanking move was attempted to support the
attack. This tactical maneuver failed to develop when the men became
bogged down in the soft sediment of the muddy marsh and Southern
reinforcements reached the rear of the battery. The disastrous Federal
assault was over after less than an hour of fighting. The Federals would
try a second assault on the battery from the north on James Island, but
it too would be compromised by the presence of the marshes and their
impassable sticky mud. During the battle, Hunter and the Union lost
more than three times as many men as the Confederate defenders, suffering more than 680 casualties. The Pleistocene ridge and adjacent
marshes had been a force multiplier for the Southern defenders, a factor summarized well by local historian Warren Ripley: "The Confederates celebrated their victory, but the real salvation of Fort Lamar lay in
mud—gooey, bottomless pluff mud—and luck."[34]

If nothing else, the attack on Fort Lamar demonstrated the folly of
attacking a sand fortification from a predictable, geologically predeter-
mined direction of approach. This was a lesson the Federal command
failed to learn, unfortunately, as eleven months later they would attempt to use essentially the identical infantry tactics on a larger scale on
a modern barrier island that was clearly visible across the salt marshes
from the reconnaissance platform in the Tower Battery: Morris Island.

Strike against Sumter

Gillmore's plan to capture Charleston was centered on reducing the
firepower the Union navy would face when attempting to enter the harbor. The navy had already tried to run the gauntlet between Batteries
Wagner and Gregg and Forts Sumter and Moultrie, with disastrous results. After the debacle on James Island, Union efforts to reduce Fort
Sumter and capture Charleston Harbor were transferred to the navy.
In the spring of 1863, the commanding officer of the South Atlantic
Blockade Squadron, Rear Adm. Samuel Francis DuPont, decided to attack the masonry defenses around the entrance to the harbor. His attack would employ the newest weapons in the fleet, his improved monitors, to steam up the harbor channel and attack Fort Sumter from a
path adjacent to Morris Island. DuPont had nine ironclads available for
the assault, and these warships carried thirty-two heavy guns of a cali-

ber larger than those arming the Southern fortifications on Morris Island or Fort Sumter.

The guns in the forts were smaller but faster to load and fire, and there were nearly four hundred pieces available for the defense. Fort Sumter alone could boast seventy-six cannon. The Confederates also had a defensive backup plan: even if the Federal ironclads were able to fight their way past the outer defenses of Charleston Harbor and survive the cross fire from Fort Sumter and Moultrie, two Confederate gunboats stood by to defend the city, the CSS *Chicora* and *Palmetto State*. Charleston Harbor also had dozens of smaller batteries lining the harbor closer to the city waterfront and many guns emplaced directly on the peninsula. This was a well-fortified city, not New Orleans.

The Federal flotilla intended to attack during the first week of April 1863, but poor weather delayed the initial assault. The timing of the offensive was critical for the underpowered ironclads: the tides in Charleston Harbor often exceeded 6 feet per second, while the top speed of the Federal vessels only ranged between 10 and 15 feet per second.[35] The attacking ships needed to remain in proper line of battle to ensure a coordinated attack, so they could not exceed the maximum velocity of the slowest vessel. A repeat of the Union naval success at New Orleans under the Confederate guns would be a slow and dangerous undertaking for the Federal ironclads against an ebb tide.

During the Civil War the Charleston Harbor ebb tidal delta contained four channels for navigation. The largest channel, the Main Ship Channel, ran from south to north, paralleling Morris Island, before turning west and passing Fort Sumter. This channel would be used for the ironclad bombardments of both Fort Sumter and Battery Wagner. Three smaller channels ran across the delta to the north of the Main Ship Channel. Shallow-draft blockade-runners were known to use both the shallower Swash Channel and North Channel. Their preferred route, however, was the channel closest to Fort Moultrie: the Beach Channel, or Maffitt's Channel.[36] All of these passages were difficult to navigate, in large part because of the constantly shifting sand. A U.S. Coast Survey map constructed from soundings taken in the 1850s indicates that the path of maximum depth through Maffitt's Channel was shifting to the north and south more than 100 feet per year. As the channel width narrowed to less than 200 feet in several locations, the treacherousness of this route for deeper-draft ships became clear.[37]

When the naval offensive started, the Federal vessels were surprised by the strong and unpredictable currents coursing across the Main Ship Channel. The *New Ironsides*, DuPont's flagship, became practically impossible to control in such currents and was anchored to avoid running aground on the shallow shoals.[38] As she did so, the ship immediately became an obstacle for the next vessel following in line.

The combination of a relatively narrow and hazardous channel, strong currents, and massive quantities of incoming enemy artillery rounds doomed the attack. When the tides began to shift, DuPont realized that any of his damaged warships were in distinct danger of being carried by the currents even closer to the enemy guns.[39] At this point he had no choice but to call off the attack and retire farther outside the harbor entrance.

During the attack the Confederates had fired 2,200 rounds with impressive accuracy. Almost 25 percent of the shells struck iron and wood, although only one Federal vessel was lost; the USS *Keokuk* sank the next day in the shallow water directly offshore from Morris Island. Her 11-inch guns would eventually be salvaged and added to the rebel defensive works around the harbor entrance. Thirty-four shots hit Fort Sumter, but only fifteen did damage; a few strikes left 2.5-foot-deep craters, but overall the fort remained in excellent condition.[40]

It was now clear that the capture of the city would not be possible using naval firepower alone. In fact, no Federal flotilla was successful in suppressing even a single Third System fortification during the entirety of the war.[41] Instead, a combined strike from the army and land-based heavy artillery, with the full support of the Federal naval guns offshore, would be required for any future successful assault against the city and its defenses.

Back to the Beach

To give the navy better odds of successfully running the cross-fire gauntlet between Moultrie and Sumter, and over and around multiple subaqueous obstacles, Gillmore proposed to use a strategy that was derivative of his successful attack plan at Savannah. Heavy land-based rifled artillery would be used to diminish Fort Sumter, shifting the firepower equation to favor the navy. With Federal ironclads in the harbor, the city that started the Civil War would have little choice but to surrender.

When formulating his plan to capture Morris Island and eventually

Charleston, Gillmore should have given more careful consideration to
the Federal folly of an earlier assault on the Pleistocene barrier island at
Secessionville. Instead, after fortifying Folly Island and striking across
Lighthouse Inlet, he would repeat the bloody barrier-island tactics on
Morris Island.[42]

Gillmore selected Brig. Gen. George Strong to lead the beachfront
assault across the inlet. Strong would command half of Gillmore's available infantry, a force in excess of five thousand men. They would use
the Folly River to move around Rat Island, eventually crossing the inlet above the ebb delta to land at the beach marking the southern inlet edge of Morris Island. All of this would be done while under fire
from Confederate artillery and, when the range closed, infantry. Suppressive fire for this exposed flotilla would be supplied by Federal artillery on Little Folly Island, the naval gunboats offshore across the ebb
tidal delta, and several Dahlgren boat howitzers.[43] The diminutive boat
howitzers would sail with the amphibious flotilla, offering close, if limited, artillery support. In total, Strong had an initial waterborne assault
force of around twenty-five hundred infantry, with a reserve on Folly Island of another twenty-eight hundred men.[44] The Federal batteries on
Little Folly Island opened fire at daybreak on the morning of July 10,
exposing their position for the first time to the Confederates a half mile
away across the inlet. The rebel batteries on Morris soon replied, and an
hour-long duel began. As shot and shell flew back and forth across the
inlet, Strong's boats remained hidden behind the reeds on Rat Island,
the northernmost marshy extension of Folly. Around 6:30 the Union
monitors *Catskill*, *Nahant*, *Montauck*, and *Weehawken* joined the fight.
The enfilading naval gunfire from offshore to the east proved devastating to the Confederate position, with grapeshot and shells tearing
along the long axis of the Southern batteries and rifle pits. Soon Confederate return fire diminished and the Federal boat howitzers entered
the fray from the opposite direction, sailing downriver toward the inlet
with Strong's infantry-filled barges following closely behind.

Strong ordered his force to land along the beach on the western end
of the Confederate entrenchments. The first wave of Federals ashore
carried new seven-shot Spencer repeating rifles, an ideal weapon for
assaulting an enemy concealed in rifle pits—assuming a soldier could
keep the mechanically complex gun away from the beach sand. Other
regiments followed closely behind, rapidly crossing the 100-yard-wide
beach. Fierce hand-to-hand combat in the trenches and rifle pits led to

FIG. 15.4. Morris Island's Battery Wagner illustrating a sea-oats covered parapet and timber palisade wall. Library of Congress.

the wounding of the local Confederate commanding officer, and the Southern defensive line soon collapsed.[45] As the primary Federal assault was proving efficacious, the Sixth Connecticut, which had landed farther toward the ocean along the inlet, began fighting across the beachfront to the north. From this position the men began attacking the vulnerable rebel gun batteries from the rear. Once the artillery positions were overwhelmed, the remaining Confederate infantry was nearly surrounded. The decimated Southerners determined that the only viable path to survival was a rapid retreat to the north through the 35-foot-high sand dunes to the protection of the heavy guns of Battery Wagner (see figure 15.4).[46]

Strong ordered his men to pursue the routed rebels and, if their momentum allowed, to seize Battery Wagner. The exhausted Federal infantry made a valiant attempt, but the combination of long-range solid shot from Fort Sumter and grapeshot and canister from Wagner ended their pursuit.[47] Certainly the heat of the day and the difficulty running across dry, fine dune sand helped ensure that the pursuit was not successful.

Gillmore's orders for the next morning, July 11, demonstrated his lack of respect for the sand fort to his front. The Ninth Maine, Seventy-Sixth Pennsylvania, and a portion of the Seventh Connecticut were ordered to attack the fort, with the assault being conducted directly along the beach. Gillmore was also apparently unaware that the Confederate garrison had been bolstered by reinforcements that had arrived on Morris Island the previous day. Strong assembled his men into columns for the attack, with four companies of the Seventh Connecticut in the lead.[48] The men were to attack into the face of the Confederate artillery and small arms fire over a narrow isthmus of sand that was bordered to the east by the Atlantic and to the west by a low marsh.[49] The Confederates had additionally deployed skirmishers in the sand ridges 100 yards in front of the battery. At dawn the next day, these pickets provided the first alert of the upcoming Federal attack.

Strong's force moved up the beach in two columns separated by the strand line. They were met with sustained Confederate fire, yet they were able to cross the battery's moat and scale the parapet. Unfortunately for the men from New England, however, the initial surge of the Seventh Connecticut was not well supported by the Ninth Maine or Seventy-Sixth Pennsylvania, and the assault faltered. The Union lost almost four hundred men in his first organized attempt to capture Battery Wagner. Nevertheless, Gillmore decided it was still conceivable, and preferable, to capture the sand fort without initiating a protracted siege.

During his next attempt to smother the Confederate defenses, Gillmore would combine a land-based cannon fusillade with gunfire from the navy's heaviest artillery to disable the Southern guns in Wagner (see figure 15.5). With the firepower of the battery diminished and the Confederate garrison wounded, his next and larger attack would certainly meet with more success. Gillmore brought forward more than thirty heavy artillery pieces to be assembled within range of the sand fort. These guns were scattered across southern Morris Island and on Black Island, a Pleistocene barrier island that the retreating Morris Island was slowly truncating. These guns would be less than a mile from the Wagner and, when the ironclads opened up, would provide a presumably deadly cross fire. This land-based firepower was minimal compared to what Gillmore had assembled on Tybee Island, at least with respect to the size of the individual pieces, but Gillmore believed the guns were more than sufficient to quiet the fort, especially when combined with the huge guns bombarding from offshore.

As Gillmore sited his artillery on the southern end of Morris Island, Confederate general William Taliaferro, the commanding officer of the Confederates on the island, was receiving additional reinforcements.

FIG. 15.5. Inspection of the Mounting guard in Fort Wagner. This photograph was taken in April 1865, and the former Battery Wagner is now in Federal hands. Library of Congress.

He now had more than 1,250 men to resist the anticipated Union attack. As he surmised, scattered ranging shots began to fall on and around his position on July 18.

By noon the Federal bombardment had commenced in full, and soon the fort's guns ceased returning fire. In a pattern that was repeated multiple times during the Civil War, this lack of counterbattery and return fire was mistakenly interpreted as a sign that the enemy's guns had been destroyed, instead of an attempt by the gun crews to conserve ammunition in anticipation of an imminent infantry attack. Earlier that month, for example, the Union artillery on Cemetery Ridge was in a similar tactical situation before the impending assault by Pickett's, Pettigrew's, and Trimble's men. The silence of the Federal guns was misinterpreted by the Southern commanders, leading to predictably devastating results for both Lee and the Army of Northern Virginia.

In Battery Wagner, nine-tenths of the batteries were still intact post-bombardment, although many of the pieces were covered in sand. The garrison remained sheltered in the fort's large bombproof. Gillmore conferred with Brig. Gen. Truman Seymour, the man who would lead

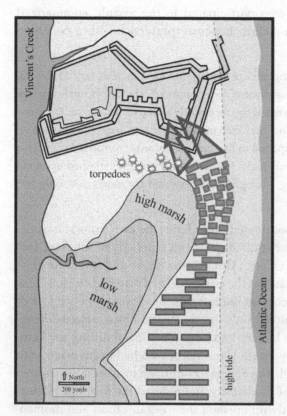

FIG. 15.6. The back-barrier marsh restricted the avenue of attack for the 54th Massachusetts during their disastrous attack on the evening of July 18, 1863. Modified from Gardner, 1891. Map from Emilio, L. 1891, *History of the Fifty-fourth Regiment of Massachusetts Volunteer Infantry, 1863–1865* (Boston: The Boston Book Company), p. 80.

his next assault, about the most favorable path for the six-thousand-man-strong assault. The generals concurred that the assault should begin at dusk, allowing just enough daylight for the men to find their way between the dunes and the waves while minimizing the visibility of the attacking force to the gunners in Wagner. The Fifty-Fourth Massachusetts, a large regiment composed of six hundred African American troops, would lead the assault. During the preceding week these men performed admirably during fighting on James Island. Land-based artillery continued to provide suppressive fire as the Federal infantry shuffled down the beachfront. The onrushing men now faced the familiar problem witnessed at Secessionville: beach sand is difficult to cross on the double-quick, and the island grew narrower just in front of the Battery, forcing the men to compress ranks and become an invitingly dense target for the Confederates (see figure 15.6). Their progress was slowed by the flanking wet sand and marsh mud, and disorganization began to become apparent in the compressing columns.

George E. Stevens, who participated in the assault, summarized the difficulties of the attack in his correspondence with a Northern newspaper:

> Moving at quick time, and preserving its formation as well as the difficult ground and narrowing way permitted, the Fifty-fourth was approaching the defile made by the easterly sweep of the marsh. Darkness was rapidly coming on, and each moment became deeper. Soon men on the flanks were compelled to fall behind, for want of room to continue in line. The centre only had a free path, and with eyes strained upon the Colonel and the flag, they pressed on toward the work, now only two hundred yards away.[50]

In a confused and uncoordinated night fight, the men from Massachusetts traversed the sediment-filled moat and climbed the southern face of the battery. Here their leader, Col. Robert Shaw, and two of the regiment's captains were killed on the parapet. The Forty-Eighth New York and Sixth Connecticut entered the fortification closer to the sea face of the battery, and they found it only lightly guarded by scattered defenders. This segment of the battery had been abandoned during the bombardment and assault by North Carolinians who had been tasked with holding the Atlantic wall. The Federals continued to ascend the parapet along the eastern side of the battery and entered the fort, but maneuvering in the dark greatly diminished their organization and momentum. As supporting Union infantry crossed the beach toward the battery, friendly fire soon began to add to the Union casualties and only intensified the confusion.

Seymour called for more reinforcements to be thrown into the assault, but Gillmore demurred. By midnight the assault was over, and what was left of the Fifty-Fourth Massachusetts entrenched at a rallying point just outside of rifle range of the fort. The second assault on Battery Wagner resulted in more than seventeen hundred casualties, only two hundred of which were Confederate. The Fifty-Fourth Massachusetts alone lost more men than the Confederate defenders in the fort.[51]

Gillmore's reaction to this ghastly defeat on the beach was to dig in, preparing to conduct a more extensive siege. On the southern end of Morris Island, a palisade was constructed in case of an unlikely Southern counterattack, and several new heavy artillery batteries were added to the Union position. It was during this time that one of the key haz-

FIG. 15.7. This photograph, probably taken in August 1863, show a remarkable perspective of Battery Hays and what is left of Fort Sumter (background, left center of the embrasure across the marsh; detail below). The 8-inch Parrott rifle has been dismounted and sits on blocks, in preparation for relocation. Detail and enhanced photograph from the Library of Congress.

ards of constructing gun batteries in dune sand became apparent. Brig. Gen. J. W. Turner describes the "serious evil" of sand:

> The material of our field works upon Morris Island was dry, hard, flinty sand, which in a windy day was constantly blowing about, and at times to such an extent did it fill the air, that it was a most severe annoyance to officers and men. On such occasions it was almost impossible to keep the pieces free from it; and at all times the sponge and rammer staves, moist from the hands of the men striking the sides and soles of the embrasures, would carry in no inconsiderable quantity. No doubt this difficulty was an extreme one with us in the position of our batteries, and I am of the opinion that it entered as an element, to some considerable extent, causing the destruction of the guns which have burst lately, though by no means sufficient in itself.[52]

Apparently the sand was better at silencing the Federal guns than were the Confederates.

Gillmore used two of these new batteries to harass the garrison in Battery Wagner, while the remainder of his artillery was concentrated on Fort Sumter (see figure 15.7). The four ironclads sailed closer to shore to join the shelling of Wagner and Sumter from a different direction. On August 17 the combined land- and water-based artillery crashed more than a thousand shells into Fort Sumter, disarming seven of the fort's remaining guns and severely breaching the brick walls.[53]

After two weeks of shelling of the brick fort, the results were beginning to resemble those at Fort Pulaski: Gillmore's heavy Parrott guns silenced any counterbattery fire and crumbled the brick parapets. By the end of the bombardment, Fort Sumter was reduced to a single tier of rubble. Concurrently with this artillery action, Federal sappers and infantry gradually dug approach trenches to Battery Wagner along the center of Morris Island. These entrenchments zigzagged their way toward the battery so that Confederate sharpshooters and artillery in the sand fort were never provided a clear shot down the length of the trench.[54]

The Federal sappers also constructed a large battery in a location that would have been thought impossible earlier in the war: in the middle of a back-barrier marsh. Sandbags and pilings were used to build a supporting platform across the mud. These pilings needed to be driven through the soft marsh sediment to reach the less compressible Pleistocene sands below, or the platform would never support the weight of the 200-pounder Parrott rifle or the gun's ammunition. This cannon, dubbed the Swamp Angel, began firing incendiary rounds at downtown Charleston within a week of completion.[55] The bombardment, albeit for only three dozen rounds before the gun exploded, marked the first deliberate targeting of a civilian population during the war.[56]

Gillmore ordered another strike against Battery Wagner after the guns of Fort Sumter could clearly no longer offer support to the sand battery. Col. George Dandy was instructed to capture the Confederate rifle pits immediately below the earthworks with the men of the 100th New York Regiment. They were momentarily successful, before a Confederate counterattack forced a withdraw. After this demonstration of the vulnerability of these forward earthworks, the Confederates began digging to improve their forward entrenched position. Before this was accomplished, however, the Southerners were again attacked and overrun by the Third New Hampshire and Twenty-Fourth Massachusetts. The Federals then began to modify and reverse the former Confederate rifle pits, and this position was converted into the closest siege line yet to the adjacent fort. This new siege line was well within rifle range of the battery and, importantly, was north of the marsh constriction that had compressed and mired the earlier infantry attacks on the earthworks. Gillmore prepared for a third assault on the fort, which he scheduled for September 7. The attack was to concentrate on the seaside sector of the battery, with supporting troops shifting around the eastern wall of

FIG. 15.8. This photograph of the flat beach along the northern half of Morris Island was taken approximately two months after Battery Wagner was abandoned. In the background (below) five ironclad monitors can be seen in the distance, along with the USS *New Ironsides*. Detail of a photograph from the Library of Congress.

the fort and attacking the weaker rear of the works. September 5 saw Gillmore and Dahlgren commence a day-and-a-half-long bombardment of the battery to (once again, it was hoped) weaken the Confederate defenses prior to attack. The effect of this bombardment was greater: Wagner's bombproof magazine began to become exposed under the shifting sand. This vulnerability, combined with the proximity of the Union siege lines, convinced the garrison commander, Col. Lawrence Keitt, that the sand fort should be abandoned once the shelling ceased. With Battery Wagner undefended, Battery Gregg, with artillery that faced toward Charleston Harbor, was untenable (see figure 15.8). As a result, the final Confederate defenders of Morris Island left Cummings Point on the evening and early morning of September 6 and 7. When Federals captured two of the Southerners' evacuation boats, Gillmore was alerted that Morris Island was now exclusively occupied by his men.

The cost of the island and battery had been extraordinary. The Union lost almost twenty-five hundred men trying to take an elaborate pile of sand. The Confederates lost only 640 men, demonstrating the force-multiplying effects of the beach and dune sand and the marsh mud, when properly exploited on the defensive.

Charleston, and the pile of debris that had been Fort Sumter, remained in Confederate hands until after Sherman had marched from Atlanta to Savannah.[57] His army would have to choose between two inviting targets in South Carolina: the state capital at Columbia, and Charleston. On Valentine's Day 1865, Beauregard decided to finally abandon the city, ordering any Confederate troops still around the city to make their way north to join Joseph Johnston's army in North Carolina. Two days later Fort Sumter was abandoned. The siege of Charleston had taken 567 days, the longest duration of any siege during the Civil War.[58]

Quincy Gillmore's operations on Morris Island are remembered for several circumstances that foreshadowed events to come during the later world wars. His tactical use of an elaborate system of entrenchments and nighttime bombardments from heavy artillery illuminated with calcium lights were predecessors to the First World War.[59] He was also one of the first commanding officers to turn his artillery, specifically the Swamp Angel, against an enemy civilian population.

Gillmore reported to Washington his astonishing findings with respect to the strength of the sedimentary forts in several reports. He acknowledged his underestimation of the fort, suggesting one source of his error: "Fort Wagner was found to be a work of the most formidable character; far more so, indeed, than the most exaggerated statements of prisoners and deserters had led us to expect. Its bomb-proof shelter, capable of containing from 1,500 to 1,600 men, remained practically intact after the most severe bombardment to which any earthwork was ever exposed. . . . The history of sieges furnishes no parallel case."[60] Gillmore reported on the futility of firing heavy artillery at sand, a waste of effort and ammunition:

> The attempt to form an opening in the bomb-proof by breaching failed for want of time. The heavy projectiles were slowly eating their way into it, although their effect was astonishingly slight. Indeed, the penetration of rifle projectiles into a sand parapet, standing at the natural slope, or approximately so, is trifling. They are almost invariably de-

flected along the line of least resistance, or departing slightly from it, scooping out in their progress a small hollow, the contents of which are scattered but a short distance. Under such circumstances, the general effect produced by firing a large number of successive shots within a small area of, say, from 15 to 20 square feet, is by no means commensurate with the expenditure of ammunition involved.[61]

Of particular note in this quotation is Gillmore's mentioning of the angle of repose for the sand. The revetment used to hold back sand, whether made of wood or sod, was the vulnerable part of a sand parapet, not the sediment itself (see figure 15.9). Walls without revetment, standing at or close to the angle of repose, are even more difficult to reduce with gunfire.

This was not the first time that Gillmore and his subordinate engineers took note of the angle of repose of the sediment. After the capture of Fort Pulaski, they were careful to report the damage done to their own sand batteries on Tybee Island.[62] Lieut. T. B. Brooks reported to Gillmore that after the Confederate shelling of their position, the "angle of slope of the parapets, traverses, sides of magazines, [etc.]" was measured at ten points and averaged 32–33°, exactly in line with the maximum angle of repose for dry, medium/fine sand.[63] These measurements suggest that the sand had suffered little displacement after the

FIG. 15.9. Naval guns on Morris Island. Lumber and sod revetments are used through most of the fortification but (re)movable sandbags are used where enemy shellfire would be more intense, near the embrasures. When hit by enemy fire the sandbags could be quickly replaced to repair the embrasures. Library of Congress.

limited Confederate counterbattery fire, with the sandy earthworks generally standing as tall as they had been when constructed. Photographs of the Confederate batteries after surrender reveals similar slopes to the sandy parapets on Morris Island.

In his report Gillmore, always an engineer, provided statistics for the accuracy of his heavy artillery fire and the massive amount of iron thrown at the battery.[64] His 100- to 300-pounder Parrott rifles fired 1,411 shells at the Wagner from between 800 and 1,900 yards away.[65] Nearly 90 percent struck the sand fort, and more than 80 percent hit the bombproof. In total he estimated that more than 150,000 pounds of iron struck the fort without breaching either the sand parapets or the bombproof.[66] Gillmore had finally learned the value of sand as a defensive force multiplier for his enemy. His next step was to convince his superiors of his new respect for sedimentary geology.

CHAPTER 16

The Strength of Sand

Military Engineering Considerations and Sedimentary Geology

A large proportion of the higher-ranking leaders and field commanders during the Civil War received formal training in military engineering. This included coursework on the proper use of terrain and topography, as well as an introduction to the siting and construction of field fortifications. Absent from the curriculum, however, was a discussion of local geology and how it might influence the positioning of troops or the possibility of constructing earthworks.[1]

The field commander who most likely had the best understanding of sedimentary geology at the beginning of the war was Union general William Rosecrans. Before the conflict began, he had studied rocks and minerals, and he could be considered an amateur geologist.[2] It is not surprising, then, that the center of his defensive line at Stones River was a particularly well-supported position with respect to geology (figure 1.3).

An array of sedimentary and sedimentary rock characteristics influenced the tactics used by soldiers in the field and the effectiveness—and even basic functionality—of their weapons. The subsequent subsection focuses primarily on the properties of sediments that most influenced combat.

SOIL AND SEDIMENT STRENGTH

Geologists and engineers have different definitions of soil. Geologists define soil as the portion of the regolith (all the material above hard bedrock) that supports the growth of rooted plants. Engineers define soil as the material on the earth's surface that can be moved without using explosives. From a military geology and engineering standpoint, the soil and its characteristics are critical for determining whether troops can dig trenches and expediently fabricate earthworks.[3] Sediments are also important when transporting and siting artillery. A 15,000-pound siege gun needs a strong, incompressible soil for movement and emplacement.

Perhaps the most important soil characteristic, from a military perspective, is texture. Soil texture is the relative proportion of sand, silt, and clay; this ratio could be roughly determined by sappers in the field. Individual sand grains can be seen with the naked eye because they are, by definition, at least ¹⁄₁₆th of a millimeter in diameter. Most beach sand along the southeastern Atlantic coastline is around ¼ millimeter to ½ millimeter in diameter (a medium sand). Silt is smaller, so individual grains cannot easily be distinguished without the use of a hand lens. Silt grains are between ¹⁄₂₅₆ and ¹⁄₁₆ millimeter in diameter. The smallest sedimentary clastic particle, clay, is the weathering product of feldspar minerals. When clay and silt are combined they form mud and lithify into a mudstone or, if layered, shale.[4]

The proportions of sand, silt, and clay will determine a soil's strength: how well it resists deformation. Deformation may include compression from having heavy artillery dragged across the earth or having a large-caliber shell penetrate an earthen parapet. Soil strength can be broken down into two components: frictional forces and cohesive forces.

Frictional forces are most important for minimizing the impact of artillery and musket/minié balls. *Friction* in a soil or sediment is related to the density, size, and shape of the sedimentary particles. Soil or sediment moisture content and the weight of overlying sediments are also important factors when determining the frictional component of soil strength.

As sand content increases in a soil, so does friction. Suppose a person attempts to slowly smash their fist into a bucket of dry sand . . . very difficult. However, consider the same situation where the bucket is filled with mud or clay: slightly easier, especially if the sediment is wet. To extend the analogy: the most difficult material to penetrate with a fist would be a dry sand made up of angular grains of quartz.[5] Rounder sand grains produce less friction and have a lower strength. The easiest sediment to penetrate would be sand or mud that has had so much water added to it that it loses all strength and begins to behave like a liquid.

In summary, the intragrain friction between sediments will determine how well a sediment resists penetration. In the field, this was an important consideration because soldiers sought shelter behind piles of sediment (parapets), and the higher the friction of the particles in the building material, the stronger the defensive works were when resisting artillery.

Cohesion in sediments is the result of electrostatic forces between fine particles. Clay is strongly cohesive, which explains the plastic behavior of the material (until saturated).[6] Coarser sediments like sand will become more cohesive if moisture is added, because the surface tension of the water will hold the grains together. However, if the sediment dries or becomes saturated, it will lose most, if not all, of its strength.

Consider, as a historical example of the importance of soil strength, the famous attack of the Fifty-Fourth Massachusetts on Morris Island, South Carolina. As the Federals approached the Confederate sand fortress at Battery Wagner, they found that dry sand was difficult and tiring to run across, but still traversable. Compacted, damp sand in the surf zone was a little more solid and less exhausting to cross. Wet sand was impossible to run across, as was the cohesive marsh mud that is both too sticky and too soft. Thus, as the regiment made their approach, the lines of infantry were restricted to the area of the beach adjacent to the surf zone; the sediment strength anywhere else on the island was too low to allow efficient and rapid passage.

Cohesion and friction will also determine the (literal) impact of an exploding shell on an earthen fortification. In an ideal defensive situation, such as with many temporary coastal fortifications, solid shot will not fully penetrate the sandy parapet, and exploding shells will penetrate partially before detonating.[7] The friction from the sand, combined with the compression of overlying sandy parapet, will slow the expanding shrapnel before it escapes into the air at a high velocity, causing casualties.[8] The impact and explosion will also be muffled to a dull thud, certainly shaking the ground but causing few injuries. Any shell damage that results from a sustained bombardment can also be rapidly repaired.[9] After multiple early and difficult coastal operations, the Union navy learned from experience: shells were not wasted in a futile attempt to reduce massive sand parapets. Instead, the range of bombardment was minimized, and the artillery was fired directly at the gun embrasures. If enough naval or land-based artillery was available for bombardment, the garrison of the fortification could be forced below ground into the sand-covered bombproof, and no return fire would be anticipated until the shelling had subsided. In this scenario the range of fire could be reduced even further, contributing to the increased accuracy needed for striking a single gun position.

During further combined naval/army operations to capture a coastal fortification, gunfire might be directed and focused again on the parapet. Shelling would begin along the portion of the fortification where the overland assault was intended to be concentrated. If enough exploding shells hit one portion of the parapet, three anticipated results would make the eventual infantry assault easier. First, the overall slope of the parapet would be lowered as sand from higher on the wall would slump toward the base. Second, the moat or ditch in front of the parapet would be filled with sand from the surrounding explosions, making it shallower and easier to traverse. Finally, if the defenders had fortified the works with torpedoes (land mines) on the beach or dunes outside of the fortifications, the concussion from bombardment or direct strikes of short-falling shot and shell might cause premature explosion or sever the electrical wires connecting the devices to the fort.

The proportion of sand, silt, and clay found in the local environments varied considerably across the landscape and between the mainland and shoreline. This variation is the result of a combination of weathering and depth and type of parent rock, local climate, and vegetation. It is also dependent on the amount of sediment transported into the region by rivers, beach drift, or wind. The battlefields surrounding Richmond and Petersburg, Virginia, for example, have soils with abundant clay because most of the underlying bedrock is feldspar-rich granite, and feldspar weathers into clay. Charleston, South Carolina, on the other hand, is underlain by thousands of feet of sandy sediment.[10] Most of the clastic particles in this region, especially closer to the shoreline, are derived from a combination of beach processes and river deposition. As a result, the sediment is sandier and clay is not as abundant.

Ternary diagrams illustrating the proportions of clay, silt, and sand are useful for differentiating soil types and characteristics and for demonstrating the variations as they relate to depositional environments (see figure 16.1).

Coastal sand fortifications like Fort Fisher and Battery Wagner proved to be incredibly durable during and after artillery bombardments, but

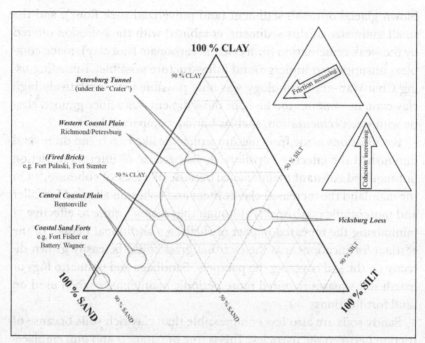

FIG. 16.1. Ternary diagram showing the estimated relative proportions of clay, silt, and sand in a local sediment and how these proportions relate to depositional environment. One nonsediment (fired brick) is added for comparison. Cohesion increases with clay content, except for the loess of Vicksburg, which is weakly cemented.

they required structural reinforcement from sandbags, palmetto logs, or squares of marsh sod because the sandy parapets lacked the cohesion offered by the intermixing of clay with the sand. Too much clay, conversely, was problematic for the Union miners when they tunneled under the Confederate earthworks at Petersburg.[11] When the strata became too clay-rich, the miners had to alter the path of their shaft to avoid the cohesive sediment because it was so exhausting to extricate.[12] When wet, clay made movement miserable. On the peninsula prior to Fair Oaks and on the Piedmont after both Fredericksburg and Chancellorsville, the Union army experienced the futility of trying to move infantry and artillery across rain-soaked, muddy Virginia roads. Nothing lowers morale more after a defeat in battle than a long slog through the mud.[13]

The loess at Vicksburg is isolated on the ternary diagram (figure 16.1) and unique in the field. The high silt content is characteristic of windblown glacial outwash sediment (and pulverized rock flour), and the small grain size of this sediment, combined with the cohesion offered by the weak cementation from calcium carbonate (not clay), made complex, unsupported underground infrastructure possible. Tunneling using Civil War–era technology was only possible with a relatively high clay content—but not too high, or the sediment was a finer-grained clastic with weak cementation, such as buried, compacted loess.

Fortifications made from mature sediments like beach and dune sand diminished the effects of artillery shells because of intergrain friction but required constant maintenance because of a lack of cohesion.[14] On the mainland the increased clay content and cohesion allowed for taller and more stable earthworks, although they weren't quite as effective at minimizing the repeated impact of shells. An additional benefit of the earthen fortifications was the fact that grass could be easily grown directly on the soil covering the parapets. Sandbags and palmetto logs or marsh sod squares required more periodic maintenance when used on sand fortifications.

Sandy soils are also less compressible than clay-rich soils because of friction between the particles. This is true of parapets and gun emplacements in forts as well. Heavy Federal artillery required minimal stabilization on Morris Island, South Carolina, when placed in the dune fields. But less than 100 yards away and in the back-barrier marsh, the Swamp Angel, an 8-inch Parrott rifle that weighed 16,500 pounds, required the construction of a foundation that included multiple pilings driven into the underlying Pleistocene "basement" sand, twenty feet below the surface, and more than thirteen thousand sandbags.[15]

The construction material of Fort McAllister on the Ogeechee River in Georgia demonstrated the effectiveness of mixing sand and mud in the right proportions for an earthen fortification. Sand diminished the effectiveness of the incoming enemy gunfire, while a smaller amount of clay provided cohesion for the structure and the amount of revetting—and maintenance—could be lessened. Seven different bombardments of the fort by the Union navy demonstrated that parapets with too much mud were not as resistant to shellfire as those made with a mixture of more sand and less mud, although both types of earthworks were equally easy to repair quickly. As a result, as the war progressed the Confederates systematically increased the sand content in the remainder of the earthworks surrounding Savannah.[16]

PENETRATION AND BLAST EFFECTS IN SEDIMENTS

Sediment and soil are effective against all three types of shelling. For direct fire, where the intent is to shoot directly at and penetrate a fortification, sand and gravels rely on friction to slow shells and diminish the depth of penetration.[17] For indirect fire, when mortars or artillery fire over a fortification and time the shell's explosion using fuses to rain fragments from above, thick earthen bombproofs proved impervious, protecting both the fortification garrison and ammunition and gunpowder supply. Finally, fragmentation effects, including primary effects from exploding shells, or secondary effects from the disintegration of both shell and target, were dramatically diminished by clastic sediments. Contemplate two soldiers taking cover in an embrasure: an enclosed gun position within a parapet. The first soldier is in a traditional brick fort, and the second is in a temporary coastal sand fort. A rifled artillery shell enters both embrasures before exploding. In the brick enclosure the soldier is exposed to fragments of iron from the exploding shell as well as fragments of brick and ricocheting iron created by the blast. The rapid increase in thermal energy and shock wave create a compression wave (overpressure in the atmosphere) that is little attenuated by the masonry. In contrast, the soldier in the sand enclosure, if fortunate, will suffer few of these effects. If the fuse was not perfectly set—which was nearly impossible in the Civil War—the shell will penetrate the sand parapet or traverse before exploding, and the resulting energy release from the blast will be confined inside the earthworks, essentially stifled. Fragments of iron will immediately be slowed by the friction encountered with the compressed sand, and any secondary projectiles will be made of sand, not angular brick fragments.

Modern military studies of the effects of shellfire on fortifications and materials provide insights into the effectiveness of sediments at dissipating explosiveness despite obvious differences in firepower and ballistics between modern and Civil War ordnance. The U.S. Air Force was interested in increasing the protection of aircraft and personnel using improved modern earthworks.[18] Their report described three types of blast on potential "soft" fieldworks: overhead burst, contact burst, and delayed fuse burst. Overhead burst from mortars or artillery is akin to indirect fire and had mixed rates of success during the Civil War. Fuses for timing of the shell explosion were not reliable, and shells often exploded short of the target or too high to cause damage. When they did explode at the proper distance and elevation, they caused little damage to well-designed and well-constructed earthen fortifications. Civil War soldiers usually preferred to wait out bombardment hidden in bombproofs, but when that was not possible, the fortifications garrison had to be reminded not to instinctually lie down during such a bombardment; exposure to explosions from above is greater for a prone soldier than one who is standing because of the difference in surface area relative to the direction of incoming shrapnel. During most coastal bombardments in the later portion of the war, Confederate soldiers usually remained in the fort's bombproof for the duration of a bombardment, trusting that any infantry assault would take place after shelling had ceased. The delay between the shelling and the subsequent infantry attack would allow time to reman and reposition or repair what was left of their artillery and parapets.

Contact burst creates damage by having a shell explode on impact, causing excavation and a ground shock that may result in the collapse of structures. Delayed fuse bursts rely on penetration of the target before the shell explodes. Sediments may cause a contact burst to behave more like a delayed fuse burst because of the initial minimal resistance to penetration.[19] Both types of bursts will be subdued in most instances when penetrating into massed sediments.

One exception to this trend that has been poorly studied with respect to energy wave and shell blast is material amplification. In earthquakes, for example, the amplitude of seismic waves is greater in unconsolidated material when compared to hard rock. This may seem counterintuitive, but a person living in San Francisco would feel less intense shaking during a major earthquake on the hard, mountainous rock region of the city than in the Marina District, which was built over unconsolidated clay- and silt-rich fill material. Shaking—and resulting damage—is always more intense on unconsolidated material.

For a large artillery explosion adjacent to an earthen fortification, the same amplification might make the resulting shaking slightly greater for a fort built atop sediments instead of one anchored on hard rock. It should be noted, however, that any massive stone or brick fortification built on unconsolidated sediment will have deep pilings to support its weight, which will, in all probability, diminish some of the impact of the material amplification.

The U.S. military has spent a great deal of time studying the pro-

TABLE 16.1. Inches of various materials needed to offer protection from an explosion 50 feet away

Material	152 mm HE shell	100 lb. bomb
Brick masonry	8	8
Brick rubble	12	18
Lumber	14	15
Soil	16	24
Wet soil	32	48
Gravel/crushed stone	12	18
Dry clay sandbag	20	30
Wet clay sandbag	40	60
Gravel/pebbles/small cobbles mixed	20	20
Dry sand sandbag	18	30
Wet sand sandbag	36	60

SOURCE: Modified from Department of the Air Force, Air Force Handbook 10–222, v. 14, 2008, *Civil Engineering Guide to Fighting Positions, Shelters, Obstacles, and Revetments*, table A.2.1.

tective characteristics of materials, including sediments and sediment mixtures.[20] Table 16.1 lists the required thickness in inches of a material needed to offer protection from a 152-mm high-explosive shell and a 100-pound bomb. Modern ballistic effects dwarf those from Civil War weapons because of new types of explosive material, increased muzzle velocity, and improved shell design. Nevertheless, insights into the protective power of sediments that would be of interest to Civil War engineers can be derived if the 152-mm shell is used as an analog for a heavy siege artillery piece and the 100-pound bomb is considered akin to a large Civil War mortar shell.[21]

Several trends are apparent from the results of this field test. Dry sediments are better for defense than wet ones when loose or in sandbags. Sand is also preferable to other grain sizes. Brick masonry and lumber appear to be ideal material until one considers the effects of repeated strikes to the same portion of a fortification. In this respect, sand and soil are more durable because the sediment retains much of its protective power even if displaced. Brick rubble offers a surprising amount of protection, which perhaps explains why the Confederate defenders inside Fort Sumter were so difficult to dislodge even after the fort had been reduced by Federal artillery.

The effectiveness of sand for protection from small arms fire is even more pronounced. Again, a modern study using high-velocity, full metal jacket rounds allows for comparison of defensive material penetration,

TABLE 16.2. Protection offered by various materials from penetration by a small-caliber machine gun bullet at 100 yards

Material	Penetration depth (inches)
Brick	18
Supported* brick rubble	12
Timber	36
Supported dry sand	**24**
Lumber	12
Supported gravel/pebble/small cobble	36
Supported dry clay	24
Supported dry loam	12
Dry clay in sandbags	40
Gravel/pebbles in sandbags	20
Dry loam in sandbags	30
Dry sand in sandbags	**20**
Parapets: clay	42
Parapets: loam	36
Parapets: sand	**24**

SOURCE: Modified from Department of the Air Force, Air Force Handbook 10–222, v. 14, 2008, *Civil Engineering Guide to Fighting Positions, Shelters, Obstacles, and Revetments*, table A.2.2.
* Listed as "structurally supported materials," i.e., there was a wall around or behind this material.

but the ballistics between a modern .30 caliber bullet and a minié ball are significant. The modern bullet has half the diameter of the earlier Civil War round but travels at approximately three times the velocity. The full metal jacket of the modern bullet also favors penetration, as the softer lead of the minié ball would not resist deformation as well when hitting sediment or rock. Nevertheless, Civil War bullets were composed entirely of dense lead, and denser projectiles are better penetrators. This is one of the reasons that tank-killing aircraft cannon often fire depleted uranium shells: uranium is even denser than lead.[22] Overall, table 16.2 from the U.S. military would underestimate the effectiveness of sand with respect to penetration by Civil War small arms fire.

Here again the benefits of friction provided by dry sand are evident. Only coarser-grained sediments and rubble are similar in terms of resistance to penetration, and these particles had the risk of becoming projectiles when hit by artillery. Of particular note is the resistance of sand parapets when compared to clay or organic-rich soil (loam).

There were also experiments regarding the material/terminal-ballistics relationship conducted during the Civil War, although they appear not to be as scientifically rigorous. The famous professor of military and civil engineering at West Point, Dennis Hart Mahan, described multiple tests on the penetration of artillery and musket fire on earthen structures in his manual *A Treatise on Field Fortifications*, republished in 1862.[23] Here he listed the depth of penetration that various types of field artillery would reach on "well-rammed earth composed of half sand and half clay" at a multitude of ranges. The deepest penetration occurred when an 18-pounder artillery piece was fired from a short range of 110 yards (6 feet of penetration). A 24-pounder cannon failed to penetrate more than 1.5 feet at 880 yards. Interestingly, Mahan noted that if sand and gravel are used in the construction of a parapet instead of sand and clay, the depth of penetration was only about 60 percent as large. He also noted that the penetration depth doubles for "common" soil that is "loosely thrown up" (not compacted).

For spherical musket balls, Mahan compared the penetration of "rammed earth, of clay and sand" to oak, boards, bundles of fascines (bound rolls of sticks and twigs), and "packed wool of mattresses."[24] These materials were fired on from ranges between 24 and 220 yards. A soldier under small arms fire could feel protected behind a solid oak fence or wall that was a mere 3 inches thick if the enemy was more than 100 yards away. At 200 yards away the fence could be as thin as 2 inches. At shorter ranges only 10 inches of soil were needed for protection, and at more than 200 yards only 6 inches of soil were required for safety. These shallow depths of penetration demonstrate the defensive value of even the crudest or most hurried breastworks. Bundles of fascines and packed wool were approximately one-third and one-fourth as effective as soil at stopping musket fire, respectively.

Finally, Mahan understood that in order to offer resilience from repeated shell strikes to a defensive work, the shell must be stopped inside the fortification's walls: "In order to insure perfect security, the thickness of parapets ought to be one-half greater than the depth of penetration furnished by experiment."[25] In other words, it was only with extra-thick parapets that a soldier would truly be protected and feel confident to stand and fight.

Quincy Gillmore communicated in his official reports about the value of sand in defensive works during the army's attempts to take Charleston in 1863. He believed that the destruction caused by his artillery fire into Battery Wagner on Morris Island was not worth the am-

TABLE 16.3. Modification of note no. 15, Maj. T. B. Brooks: "Penetration of Rifled Musket-Balls into various kinds of siege materials, ascertained by practice on Morris Island. The Sharps, Enfield, and Spencer rifles were used at distances of 10–15 yards."

Material	Penetration depth (inches)
Dry yellow pine	2.5 to 3.5
Green palmetto	7.5 to 8.5
Dry fascine	8.5 to 9.5
Dry sand in bags	**6 to 7**
Wet sand in bags	7.5 to 8.5
Loose, damp sand	8 to 14
Cotton packed in sand bags	22

munition that was consumed: his big guns did very little damage to the sand-covered stronghold, and the bombproof, hidden under a thick layer of sediment, proved especially resistant to bombardment.[26]

In total, Gillmore estimated his artillery fired more than fourteen hundred shells weighing over 75 tons into Battery Wagner, but the resulting damage to the fortification and garrison was negligible.[27] In hindsight, the subsequent disastrous attack by Union infantry across the Morris Island beachfront facing Battery Wagner was entirely predictable.

After the rebels abandoned Battery Wagner, Gillmore was able to explore the effects of his artillery firsthand. His "personal examination" of the shell damage led him to estimate that only 165 cubic yards of sand had been removed or displaced by shellfire around the bombproof to a distance and position where it would no longer offer protection. To actually "effect" or partially breach the bombproof, the general estimated, would require the direct impact of 54.5 gross tons (122,080 pounds) of metal (shells)![28]

After the underestimation of the strength of sand fortifications had been so sadly demonstrated, the Union army conducted its own penetration tests of small arms fire in the field. The durability and effectiveness of Morris Island beach and dune sand was reported by Maj. T. B. Brooks in a note from the *Official Records* (see table 16.3).[29]

Although far from rigorously scientific, sand and sandbags were demonstrated to be highly effective at slowing small arms projectiles.[30] From an engineering standpoint, Brooks additionally commented that a cubic foot of "dry, fine, light-colored quartz sand" weighs 86 pounds and will "absorb" approximately three gallons of water.[31]

Experiments both in the laboratory and the field have proven that the friction provided by beach- and dune-size grains of siliceous sand are efficient energy absorbers for both artillery and rifle musket rounds. The effect was quickly realized by Union general Gillmore on Morris Island through observation, experience, and experimentation, and his recommendation to his superiors in the Union leadership hierarchy promoted the exploitation of this characteristic of the sediment: "A comparison of the two sieges of Fort Pulaski and Fort Wagner, the former a casemated brickwork, and the latter a sand fort improvised for the occasion, leads to the query whether all of our batteries should not be constructed of a material like sand."[32]

CHAPTER 17

Gibraltar of the South

King of All Sandcastles

When Fort Gaines surrendered on August 8, 1864, and Fort Morgan followed two weeks later, the fate of Mobile, Alabama, was sealed. Farragut's damned victory closed the last remaining Confederate port east of the Mississippi on the Gulf coast, freeing up the navy to concentrate on halting Confederate international trade across the Atlantic entirely.

Only a single seaport remained open to Confederate blockade-runners along the Atlantic coast. There was a reason Wilmington was the last port to fall: the mouth of the Cape Fear River was incredibly well guarded by a combination of brick and sand fortifications, including Fort Fisher—arguably the strongest coastal fortification on earth at the time.

Sedimentary geology provided benefits for both the Confederate blockade-runners plying the Cape Fear River and the ships' defenders on land. Two inlets were available to the oceangoing ships. Old Inlet, also referred to as the Western Bar Inlet, aligned perfectly with the trend of the Cape Fear. This 17-foot-deep inlet separated Smith's Island (also called Bald Head Island) from Smithville and the mainland.[1] Smith's Island is a remnant of the Cape Fear ebb tidal delta, and the low bare-beach island was formed from the interaction of river-borne sediment and beach drift from the north.[2] The outer defenses of Wilmington are 25 miles to the north up the Cape Fear River, and a ship could sail to the inlet from Nassau or Bermuda in only two or three days, respectively.[3]

The second inlet option for Southern merchant ships was New Inlet, a shallower passage to the north of Smith's Island. This younger inlet was formed violently in a hurricane that struck in September 1761. This storm cut a path through a spit called the "Haul-over" by locals, and beach drift from the north added sediment to decrease the depth of the already dangerous navigation channel. As a result, the inlet remained around only 12 feet deep (and less than 8 feet at low tide), restricting passage to only shallow-draft ships.

Only seven miles separated the locations where the two inlets joined the sea, and Confederate shore artillery kept the Union blockade ships farther offshore. Unfortunately for the Federal navy, however, sedimentary geology essentially doubled the number of ships needed to blockade the inlets and Wilmington because of the presence of a massive subaqueous sand body called Frying Pan Shoals.

The beaches at Fort Fisher and Kure Beach are somewhat unusual in comparison to those found elsewhere along the North Carolina shoreline. For most Carolina beaches, hard rocks are not to be found. If they are present anywhere in the vicinity, they are usually located offshore, often creating submarine headlands. Northern Topsail Island and Southern Onslow Beach are two examples of beaches with sloping hard bottoms created from Oligocene limestone outcrops.[4] At Fort Fisher, in contrast, Quaternary coquina crops out directly on the beach. These rocks, composed of cemented fragments of calcareous mollusk shells, help stabilize the shoreface in some areas. Nevertheless, Fort Fisher was constructed on a portion of the shoreline that is underlain by relict stream sediments that are especially vulnerable to wave erosion. Before armoring, the shoreline at the fort was eroding at an astounding 8 feet per year, and much of this sediment was being transported by waves and currents to the south where it was added to Frying Pan Shoals.[5]

Frying Pan Shoals is a major component of the Graveyard of the Atlantic. The shallow, shifting sands produce unpredictable currents and rough waters. The shoals extend offshore nearly 30 miles, so any blockade squadron hoping to close Cape Fear's two inlets would, in effect, need to be divided into two squadrons to surround both inlets (see figure 17.1).[6] A mid-nineteenth-century navigation chart warned: "These channels over the shoals should not be attempted by strangers in vessels drawing more than 7 feet."[7]

Thus, the Union ships available for patrolling this sector of the coastline were divided into two task forces. The First Division of the North At-

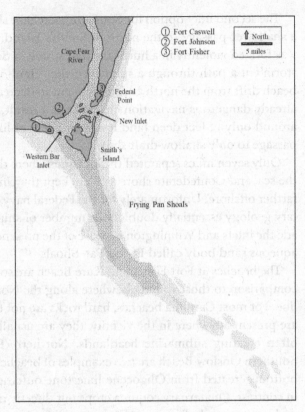

FIG. 17.1. Frying Pan Shoals represented an obstacle to the North Atlantic Blockade Squadron that required the Navy to use twice as many ships as first thought necessary: The Third Division would blockade the Western Bar Inlet and the First Division would guard the New Inlet.

lantic Blockade Squadron guarded New Inlet, attempting to minimize transport from Bermuda and points to the north. The Third Division blockaded the Western Bar route, restricting passage from the Bahamas or points to the south. The two divisions could not efficiently support each other, unfortunately, because of the presence of the unpredictable shoals. The navy would need to double the commitment of ships to close Cape Fear, with squadrons that were deceptively far apart.

The Confederates did the most with their geology-enhanced defensive scheme. Three alliterative factors allowed for a rapid construction of a series of defensive fortifications to protect the city and inlets: sand, sod, and slaves. By 1864 only Charleston had a similar defensive strength along the shoreline.[8]

The Western Bar Inlet was guarded by a series of batteries and forts, the strongest of which was Fort Caswell. Constructed between 1826 and 1838, this fort was a brick pentagon surrounded by a moat. During the Civil War the works were strengthened defensively with the addition of sand and pine logs to the front of the casemates.[9] By 1864 Fort Caswell was surrounded by a massive sand rampart, clearly a sign that the Confederate engineers had learned something from the surrender of Fort Macon and the destruction of Fort Pulaski.[10] Caswell had fifteen large artillery pieces, including an imported British 8-inch, 150-pounder Armstrong gun.

Across the inlet the Southerners built another fortification, the sand Fort Holmes. This fort was started in the fall of 1863 and was never finished, although even when incomplete it held eleven heavy artillery pieces. Despite being under construction and incompletely armed, the fortification still provided cross fire with Fort Caswell across the Western Bar Inlet. Perhaps the most important contribution of Fort Holmes was the sandwork's location: it prohibited the Federals from landing on Smith's Island, from which they might launch an overland campaign against the weaker backside of Fort Fisher.

New Inlet was even better defended than the older inlet to the south. Several sand batteries were strategically located on the beaches of Federal Point.[11] The guns in these batteries occasionally traded shots with Union vessels that approached the inlet from the north, and more than once the Confederate batteries were needed to provide cover fire for stranded blockade-runners who had been driven ashore.

By far the strongest fortification guarding New Inlet was the imposing Fort Fisher. In 1861 the fort was nothing more than a string of connected, and relatively weak, sand batteries. Several commanding officers made small improvements to the work until Col. William Lamb, along with labor from five hundred slaves, vastly expanded and improved the fort. These modifications included adding new sand parapets and batteries and strengthening the preexisting batteries.

Lamb's efforts demonstrated his respect for the Federal 15-inch ordnance, and Fort Fisher would prove to be the best example from the Civil War of the use of sedimentary geology as a force multiplier for a defensive work.[12] Anchored near outcropping coquina, the fort incorporated massive sand ramparts and immense piles of sand to elevate heavy cannon. The fort's land face crossed Federal Point, running from a largely impassable muddy marsh adjacent to the Cape Fear River across the width of the island to the Atlantic Ocean. The land face and sea face met at a full sand bastion, 32 feet high, marking the northeastern corner of the fort. The sand parapet continued to the south, paralleling the shoreline approximately 75 yards from the high water line.

FIG. 17.2. The Mound Battery, also known as Battery Lamb, held two heavy artillery pieces, one of which can just be seen on the top of the sand hill above the left side of the larger wooden shed. Photograph enhanced and modified from the Library of Congress.

The sea face ran for 1,250 yards and was anchored by two giant sand batteries.[13] Battery Meade was a crescent-shaped mound nicknamed "The Pulpit." During combat this battery served as the fort's headquarters, in large part because of its commanding view of both wings of the fortification.[14] A bombproof and hospital were located inside the sand mound.

At the far southern end of the parapet was an even higher sand mound: Battery Lamb. This battery was 43 feet tall and round, so it was dubbed "The Mound Battery."[15] From this elevation, the 10-inch Columbiad and 6.4-inch Brooke smoothbore atop the battery could send plunging fire into the inlet, attacking the vulnerable decks of approaching Union vessels.[16] The mound of quartz sand was tall enough to be seen for miles, providing a navigation aid for blockade-runners (see figure 17.2).

The construction of the Mound Battery took full advantage of the internal friction of the sand and the dry sediment's angle of repose. When construction began in the spring of 1863, a steam engine was used to pull carts of sand to the top of the growing mound. The sand was dumped at the crest, forming a massive cone-shaped artificial dune. The slope of the dune was controlled by the angle of repose—no re-

FIG. 17.3. This photograph, captioned "Southeast end showing main magazine," reveals several interesting features on the interior of the fort. Both timber and sandbag revetments are used in the batteries, and sandbags protect the entrance to the magazine (center). Detail of an image from the Library of Congress.

vetment was used—and the internal friction of the sand allowed the mound to support more than one piece of heavy ordnance.

As a final protective measure for the inlet, another massive, fortified pile of sand was constructed to the southwest, overlooking the Cape Fear River and the western extension of the inlet. This was Battery Buchanan, an oval-shaped earthwork with two 11-inch Brooke rifles and two 10-inch Columbiads. There was a landing dock adjacent to this mound of sand, and Lamb planned to use this battery in case of a last stand—a castle keep, so to speak.[17]

Because of the ease of excavation of the sand, and the vast supply of it, other defensive enhancements were added to the fortress. The artillery compartments were separated by sand traverses, protecting the guns from enfilade fire from the ocean. Even if a gun exploded, a not uncommon occurrence, only a single gun crew would be exposed to flying iron.

Under each traverse, buried beneath several feet of unconsolidated sediment, was a substantial bunker. These protected spaces provided a bombproof for the fort's defenders while the garrison was being bombarded and allowed space for gunpowder and shells in proximity to the guns (see figure 17.3).

Ease of excavation of the sediment also allowed for other, subtler improvements (see tables 17.1–17.3). Small pits were dug in a roughly linear traverse in front of, and parallel to, the land face of the fort. Torpedoes were placed into these shallow pits and buried with sand.[18] After

TABLE 17.1. Summary of defensive *advantages* for temporary coastal fortifications offered by sand

---

(1) Sand is abundant and easy to excavate and move, especially when compared to hard rock. As a result, a fort like Fisher could be quickly built with 20- to 25-foot-high parapets that were more than 20 feet thick.* On Federal Point in 1863, wheelbarrows were more important than rifles.

As an example of how quickly sand batteries can be constructed, on Folly Island, South Carolina, Federal gun placements for 32 cannon and 15 mortars were built between June 17 and 26, 1863.

(2) Sand is reusable. The sand excavated from a moat or ditch can be used to construct a parapet or glacis.

(3) Sand deflects projectiles and diminishes the effects of explosions. Sediment will also often fall back into place after being hit by artillery. Sand reportedly causes many types of shell to either ricochet or fail to explode. One after-battle report from Morris Island stated: "The slightest cause, it was found, would deflect a rifle projectile when striking upon earth or sand, and when deflected it almost invariably failed to explode."†

(4) Friction between sand grains limits penetration of solid shot. Quincy Gillmore remarked on this quality on Morris Island: "Compact sand, upon which the heaviest projectiles produce but little effect, and in which damages could be easily and speedily repaired."‡

(5) Sand bags are easy to make and move and, most importantly, bulletproof. The empty cloth bags are lightweight and easy to transport to the area where they will be filled and employed.

(6) Sand isn't flammable.

(7) Sand is found adjacent to the ocean: it can be easily piled along low-elevation shorefronts to provide artillery positions at a variety of elevations. In Fort Fisher, for example, Confederate guns were positioned in the sea face close to sea level to exploit ricocheting fire across the water. Other artillery was placed on sand mounds more than 40 feet in the air to take advantage of plunging fire onto the vulnerable wooden decks of the enemy ships.

(8) Sand structures can be modified quickly after a fortification is completed. Fort Fisher, as an example, was enlarged and changed multiple times between 1862 and 1864. The brick Fort Sumter, in contrast, changed little between construction and reduction.

(9) Sand limits ricocheting fire within embrasures, especially with small-arms fire.

(10) Sand, when hit with a heavy artillery shell, will not splinter or fracture, creating the type of deadly secondary projectiles associated with hard rock or brick.

(11) Sand is difficult for attackers to walk over and fatiguing to run over. At Fort Fisher the attacking navy sailors and marines were exhausted by the time they reached the fort's palisade wall.

(12) Clean (i.e., mature) sand is porous and permeable. It drains easily, and if saturated, sand is permeable enough to easily extract potable groundwater without significant pressure.

(13) Friction between sand grains means that the sediment can support a great deal of weight without deforming, an important feature for the site of a heavy artillery piece. Compare, for example, the extensive engineering that was necessary for the construction of the "Swamp Angel" battery in the back-barrier marsh of Morris Island, compared with the minimal preparation necessary for the heavy guns placed in the dune fields 200 yards away.

---

* Trotter WR, 1989, *Ironclads and Columbiads* (Winston-Salem, N.C.: John F. Blair), 456p.
† Gillmore Q, 1890, *The War of the Rebellion: A Compilation of the Official Records of the Union and Confederate Armies*, ser. 1, v. 28, pt. 1, p. 152.
‡ Gillmore, 1890, p. 44.

TABLE 17.2. Summary of defensive *disadvantages* for temporary coastal fortifications offered by sand

(1) Sand is not cohesive. Some form of revetment is necessary (timber, marsh sod) if sediment is piled higher than the angle of repose. As a result, rapidly dug trenches through sand will quickly collapse and fill in if not supported.

(2) Sand is porous and permeable, so it can be a challenge to keep water away from gunpowder. The torpedoes in front of Fort Fisher and Battery Wagner, buried in medium beach and dune sand, were especially vulnerable to water. This permeability also makes it difficult to contain water in moats in sandy environments.

(3) Sand gets into everything--shoes, guns, mouths, cannon. A few unfortunately situated grains of sand can render a rifle as useless as a club, and sand in the bore of an artillery piece will rapidly destroy the rifling and may cause a premature explosion.

(4) Sand bags have a limited lifespan in harsh, coastal environments.

(5) Sand can be relatively easily (and unintentionally) transported by wind and water. Heavy precipitation or windy conditions can reform sandy earthworks that aren't properly reinforced or revetted. Required maintenance can be higher in sandy earthworks.

(6) Sand, when disturbed, can temporarily have a different color. When Union forces were attempting to covertly construct gun batteries on Folly Island, the presence of darker (wetter) sand made their efforts obvious until they began to sprinkle dry dune sand on the new construction.*

* Wise SR, 1994, *Gate of Hell: Campaign for Charleston Harbor, 1863* (Columbia: University of South Carolina Press), 312p.

TABLE 17.3. Summary of the defensive *advantages* for temporary coastal fortifications offered by marshes and mud

(1) Marshes, especially low marshes, provide flank protection because they are impossible to cross for infantry—the sediment is simply too muddy.

(2) Marshes are flat, providing excellent sight lines for artillery and small arms.

(3) Mud, when combined with sand, adds cohesion to the sediment. Parapet walls or trench walls will not need to be revetted as extensively in cohesive sediment. Trenches dug through sand/mud combinations will be less likely to quickly collapse.

(4) Marsh sod makes an excellent revetment material for sandy parapets and is found nearby coastal batteries in the back-barrier marshes.

(5) Dark, adhesive mud can be used to camouflage a face or the side of a ship, rendering them more difficult to see during night operations.

burial, the beach and dune sand showed virtually no signs of having been disturbed, unlike shallow pits dug into soil. The torpedoes were connected to the fort by wires so they could be detonated remotely in the event of a land assault. A 9-foot-tall palisade of sharpened pine was added at the base of the outer parapet wall, dug into a 3-foot trench (see figure 17.4). This timber barrier contained loopholes for rifles.

The sand fort could be easily modified and improved in ways that would have been vastly more difficult had the engineers been working with a masonry structure.[19] For example, in the fall of 1864 a tunnel was cut directly through the middle of the land face parapet so that sharp-

FIG. 17.4. A portion of the massive landface of Fort Fisher after surrender. The now-damaged palisade wall runs along the base of the ramparts. Detail of a photograph from the Library of Congress.

shooters and field artillery could exit the main fort and take a position on a small, slightly elevated demilune.[20] From this position they could direct fire down the entire length of the land face, or across the half-mile-wide, flat, open plain in front of the fort. Lamb had the maritime forest and scrub removed from anywhere near the works to deny any cover to an approaching enemy.

THE FAILURE OF THE FIRST FEDERAL FORT FISHER CAMPAIGN

The importance of Forts Caswell and Fisher, as well as the commerce traveling on the Cape Fear River, were clearly identified by the leadership from both sides of the war. "Father Neptune," Secretary of the Navy Gideon Welles, put it simply: "Could we seize the forts at the entrance of Cape Fear and close the illicit traffic, it would be almost as important as the capture of Richmond on the fate of the Rebels."[21] Robert E. Lee echoed a similar sentiment, telling Fort Fisher's commanding officer that the Army of Northern Virginia would no longer be a viable fighting force without supplies from Wilmington's blockade-runners—and without Lee's army, Richmond might as well be in Federal possession.

Enter Maj. Gen. Quincy Gillmore again. Secretary Welles had gained permission to move against Wilmington's port from Lincoln and Grant, who were anxious to see Lee's supply lines diminished, but he lacked a coherent plan for attacking the sand-fortified bastions. Gillmore's reputation gained in South Carolina and Georgia suggested he had the experience to devise a strategy for the capture of the inlets, and during the first week of September 1864, he presented two potential options to the secretary and General Grant.

His first proposed plan mirrored his earlier successful design for destroying Fort Sumter: land on an undefended island to the south and move north, taking full advantage of the combination of land-based and naval artillery. This strategy called for six thousand troops to land on Smith's and Zeke's Islands to establish a beachhead. The navy would then suppress Fort Fisher with a heavy bombardment while shallow-draft monitors occupied New Inlet.

Gillmore's alternate plan incorporated the lessons learned on Morris Island regarding trenches and siege operations. A small infantry contingent would feint a landing on Smith's Island, distracting the Fort Fisher garrison, while twelve thousand infantry landed to the north of the fort and inlet on Federal Point. These men would rapidly dig entrenchments across the island, and the naval bombardment/shallow-draft inlet invasion would commence. Welles and Grant favored this plan, as it provided more flexibility for the army; if the opportunity arose, the infantry could march north and attack the city of Wilmington directly. In the worst-case scenario, the army could adopt a siege strategy similar to the one that had forced the abandonment of Battery Wagner a year earlier.

The hero of Mobile Bay, David Farragut, was selected to lead the flotilla, but he declined, citing health problems. Instead, David Porter was brought east to join with Maj. Gen. Benjamin Butler, commander of the sixty-five hundred troops designated for the assault.

Butler had a penchant for embracing particularly unusual tactics. At Bermuda Hundred, southeast of Petersburg, he attempted to dig a massive canal across a meander on the James, all to bypass a series of Confederate river batteries. Dutch Gap Canal turned out to be a promising gambit, but one that eventually proved worthless and costly in terms of manpower.

Not far from the Dutch Gap Canal project was City Point, the supply base of Grant's army around Petersburg. On August 9, 1864, the ordnance barge *J. E. Kendrick* exploded, creating a massive shock wave and spraying shell fragments and debris across the supply depot. After hearing of the scale of the destruction, Butler became infatuated with the concept of breaching the walls of a fortification with such a ship-laden bomb. Fort Fisher seemed the perfect target. Admiral Porter, never a fan of Butler's, unexpectedly agreed to the plan, in large part because he lacked a good reason not to.

Butler's plan appeared to be remarkably simple in execution: an old, flat-bottomed boat named *Louisiana* was filled with more than 200 tons of gunpowder and anything else that might add to a massive conflagration, and the ship was to be towed to the sea face of the fort and detonated.

Much was expected of this blast; Porter told his subordinates that "Houses in Wilmington will tumble to the ground."[22] Out of caution he also ordered his ships to retire twelve miles offshore to avoid any of the forthcoming blast damage.

The first campaign to take Fort Fisher was plagued by a lack of coordination between the army and navy from the start. Butler arrived offshore from Federal Point first, on December 15. Porter and his 150 ships arrived three days later, finding that Butler's transport ships had sailed back to Beaufort, North Carolina. Butler intended to return on Christmas Eve. The night before Butler's men were scheduled to return, however, Porter set the ship-bomb plan into action.

The crew of the USS *Wilderness* towed the *Louisiana* to a spot just north of the fort's bastion, or so they thought. The especially dark night, combined with a treacherous approach through shoals and sunken blockade-runners, meant the heavy and ungainly ship-bomb was nowhere near the perimeter of the fort. Instead, the boat sat almost 500 yards away from the ramparts. Making matters worse, a strong undertow current began to slowly pull the *Louisiana* offshore.

Despite these setbacks, an elaborate timed-fuse network of detonations was triggered, and a fire was started on the *Louisiana* to prevent boarding (and defusing) by the Confederates. The gunpowder in the ship was intended to be exploded simultaneously, creating the largest possible eruption and shock wave. Unfortunately, the fuse system failed completely, and the powder was ignited twenty-two minutes later by the previously set fire, creating a piecemeal detonation.

Speaking of "twenty-two": Forty years ago my father dropped a box of .22 rimfire ammunition on the hardwood floor of his home office. One of the shells hit on the primer, setting it off. The bullet traveled all of a few inches from the case, stopping harmlessly because of gravity and the friction with the floor. Without a gun chamber or other solid object to contain or direct the small explosion, the bullet and shell casing more or less cancelled each other's momentum, producing a far-from-violent result. The same thing happened with the slowly developing explosion of the *Louisiana*: because it wasn't adjacent to the sand fort, the explosive concussion would travel in all directions, minimizing the shock and damage to anything nearby.

In all, more than three-fourths of the powder failed to explode because it was either defective or saturated with seawater. Inside Fort Fisher, the blast could be heard but barely felt. This wasn't a surprise because by the time of the explosion, the *Louisiana* had drifted almost half a mile from the fort and, importantly, was no longer resting on the ocean floor. Without touching bottom, any potential seismic waves or material amplification from the sand, which might actually damage Fisher's sedimentary walls, were rapidly attenuated.

Despite this tremendous disappointment, Porter was determined to begin his naval bombardment the next morning. He deployed the six hundred guns of his flotilla in a long semicircle around the sea face of the fort and opened fire just before noon. Five hours and ten thousand rounds later, the ships moved offshore. The next day they returned and fired eleven thousand more shells into the sand fortification. The Fort Fisher garrison, wishing to save ammunition and concerned about the long range of the engagement, fired fewer than seven hundred rounds in return.[23]

At the end of the largest bombardment to date during the war, the sand fort was largely unscathed and only twenty-three Confederates had been killed or wounded. Only two guns had been (temporarily) disabled. Porter, meanwhile, was convinced the fort had been annihilated and told Butler, who was just arriving on the site, of his supposed triumph.

Butler's transports arrived in full strength the next day, and Porter's gunboats provided covering fire as the infantry disembarked on the beach to the north of the fort. Two thousand men under the command of Godfrey Weitzel landed and began to move south toward Fort Fisher's land face. They were met with the full effect of the fort's nominally diminished firepower: eighteen guns pointing directly down the beach. The large, sandy traverses had protected the gun emplacements while the gun crews hid, protected in the parapet's deeply buried bombproofs.

In the face of this cannonade and a new threat to the rear—six thousand newly arrived Confederate reinforcements from Wilmington—Weitzel halted. Butler ordered him back to the beach. From here small boats cautiously returned his men to the transport ships, and they slowly proceeded north to Hampton Roads. Weitzel had lost only sixteen men in the expedition, and Porter was livid with the decision to withdraw. The transfer of the infantry back to the transport ships meant that the navy had accomplished nothing during the operation, other than wasting a colossal amount of ammunition. Secretary Welles and General Grant concurred with Porter's opinion, and the naval commander was directed to stay in Beaufort in anticipation of a second attack by a larger force of infantry under new, presumably bolder, field leadership.

After this delay, the need to quickly capture Wilmington intensified, but not because of Southern blockade-running success. Sherman had captured Savannah by this time and was moving across the Carolinas. Wilmington was needed as a supply depot for his fast-moving army, and Grant was determined to provide this logistical support.

CLOSING THE RANGE:
THE SECOND STRIKE AGAINST FISHER

On January 8, 1865, eight thousand Federal infantry boarded transports at Bermuda Hundred in Virginia. A significant fraction of these men had formerly been commanded by Benjamin Butler. Now, however, their leader would be Maj. Gen. Alfred Terry, an officer who had served well on Morris Island during the Charleston Campaign. As an added benefit, Terry—unlike Butler—worked well with Porter.

Lamb and the Fort Fisher garrison acted quickly to repair the bombardment damage from December (see figure 17.5), and the commander requested reinforcements from his superior officer, Braxton Bragg. Bragg declined the appeal, stating that the six thousand troops available were needed to guard the city of Wilmington directly and that a second strike against the fort was not imminent. As a result, when Union forces arrived offshore of Fort Fisher on January 12, Lamb had at most eight hundred men in his newly refurbished fortification. When Porter's ships opened fire the next day, emergency reinforcements were hastily transported downriver, doubling the size of his garrison.

Porter's tactics for a second massive bombardment of the fort changed significantly. This time his ships closed the range and sent concentrated enfilading fire toward the land face of the fort. The gunfire was relentless on the 13th and 14th, prohibiting even rapid or cursory repair efforts to the sand works.[24] Robert Watson, a Confederate sailor

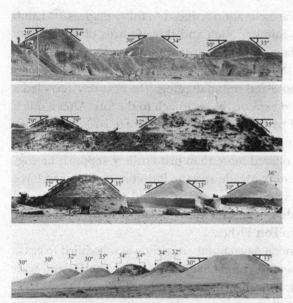

FIG. 17.5. The slopes of the parapets and traverses in Fort Fisher were consistently around the angle of repose for fine sand. Two trends can be observed when studying the slopes of these traverses: First, the angle of repose is lower towards the ocean (~30°), presumably because of displacement of the sand by naval shellfire. On the protected interior, the slope is higher (~33–35°). Second, in areas where vegetation remains post-bombardment, the slope is higher (~32–34°). The lowest slopes are consistently on the side of the earthwork facing the ocean (naval bombardment), on fresh sand surfaces. All four photographs are details from the Library of Congress.

positioned along Fisher's sea face, wrote of the impact of heavy artillery on the sand around him: "Several of us were knocked down with sand bags. . . . We were all nearly buried in sand several times. This was caused by shells bursting in the sand. Whenever one would strike near us in the sand it would throw the sand over us by the cartload."[25]

The second severe shelling had significant detrimental effects to the land face of the fort, battering the parapet and traverses. Chief engineer C. B. Comstock noted the apparent decrease in slope of the sandy parapets after being bombarded repeatedly: "The slopes of the works appear to have been generally revetted with marsh sod, or covered with grass, and to have had an inclination of forty-five degrees or a little less. On those slopes exposed to fire the revetment or grassing has been entirely destroyed, and the inclination reduced to thirty degrees."[26] Lamb estimated he lost two hundred men and three-fourths of his artillery firepower in the land face during this intense shelling.[27]

Concurrently with the naval bombardment, Terry's eight thousand men were brought ashore and began digging. The next day Terry led a reconnaissance south to explore the approach to the fort. After a quick survey of the potential paths of assault, he planned to strike the following day after consulting with Porter to coordinate the bombardment and incursion. Porter offered more than just artillery support; he suggested an attack on Battery Meade and the Pulpit by a combined force of two thousand sailors and marines. Terry would attack the other end of the land face, dividing whatever Confederate defensive fire might still remain from within Fort Fisher.

At 2:00 p.m. the general naval bombardment was scheduled to cease and the land assault would commence. At this time the Federal warships would shift suppressive fire from the land face toward the sea face, to avoid inflicting friendly casualties on the attacking infantry, sailors, and marines.

At half past two the naval bombardment paused, and rebel lookouts in Fort Fisher alerted the sheltering garrison to return to their assigned positions from the bombproofs. The sailors, unfamiliar with fighting on dry land and armed only with pistols and sabers, began to move forward along the shoreline and were rapidly cut down.[28] They quickly retreated, leaving three hundred men wounded or dying on the beach. Just as Lamb's men repulsed this futile attack, the main assault from Terry's men began. This attack had been delayed when the Confederate gunboat *Chickamauga*, which had been patrolling the Cape Fear River, discovered the force and opened fire.

Terry's assault concentrated on the three westernmost traverses of the fort adjacent to the river and marsh. Once a bloody foothold was taken on these sand mounds, the Federal infantry fought to capture the remainder of the land face before shifting the attack to the northernmost gun positions along the sea face. Several of Porter's gunboats came close to the shore and sent highly accurate shellfire into the Confederate positions, softening up the next subsequent embrasure to be attacked by the infantry. This preliminary tempering of the enemy's strength reduced the bloodshed on the part of Terry's men.

Once the land face and northeastern bastion were in Union hands, the sea face fell fairly quickly (see figure 17.6). By 10:00 p.m., Maj. Gen.

FIG. 17.6. Timothy H. O'Sullivan captured this image of the interior of the seaface of Fort Fisher during February 1865. The gun in the center has been disabled by the Union bombardment, while the cannon on the left and right have presumably been reversed to fire at the attacking Federal infantry across the parade ground and along the interior of the landface. Modified and enhanced photograph from the Library of Congress.

James Reilly, the highest remaining unwounded Confederate officer, surrendered what was left of the fort. The assault on the Gibraltar of the South cost the Union navy 383 men, and the army lost 955 soldiers. The overall Confederate casualties were similar: the entire fifteen-hundred-man garrison was lost to surrender, with around five hundred of these men being either killed or wounded.[29]

After the victory the fort would take another 100+ lives that the sand would do nothing to prevent. Ill-fated soldiers from New York and New Hampshire set up camp directly above the fort's primary subterranean magazine, adjacent to the parade grounds. Apparently, two curious soldiers who had celebrated a bit too much wandered into the dark magazine under the camp with torches, igniting 13,000 pounds of gunpowder. In the resulting terrific explosion, men were thrown as high as 600 feet in the air, and others were killed when they were buried in the sand and suffocated. Still others found themselves suddenly sinking below the surface of the sand when the sensitive sediment lost all strength during the violent shaking. The forty-year-old surgeon of the 117th New York, James A. Mowris, was on the exterior slope of the fortress's parapet between the northeast bastion and the Pulpit when the explosion occurred. He was knocked to the ground by the seismic waves and found himself buried in a fresh sand grave with a broken arm.[30]

In the end, the fall of Fort Fisher would represent the power of a coordinated army/navy assault, and both Porter and Terry would speak nothing but great compliments of their counterpart. The surrender of the fort would also (again) demonstrate the poor leadership of Braxton Bragg, who somehow overestimated the strength of the fortification and garrison and underestimated the firepower available to the Union. His refusal to provide proper support to or reinforcements for the beleaguered garrison directly led to the capture of the fort.

Fort Fisher had one defensive vulnerability that was not present at Battery Wagner on Morris Island. Both fortresses had a land face anchored above a flanking marsh, but on Morris Island a tidal stream, Vincent's Creek, made passage around the end of the battery impossible. At Fort Fisher, in contrast, Shepherd's Battery rose above a wide, flat, muddy marsh. Coastal marshes are divided spatially by a number of zones of elevation relative to sea level: low marsh, intermediate marsh, and high marsh. Each zone has a distinct fauna and flora, and importantly, each has a different sediment type. Low marshes are muddy, composed of wet clay and silt, with little sand. High marshes are sandy, with a lower component of silt and clay. This sedimentary zoning is especially apparent in the marshes between the Cape Fear River and Fort Fisher.[31]

Along the Cape Fear the low marsh is found closest to the river, and the high marsh runs along the fringe of the mainland. As a result, when Abbott's Brigade attacked the wounded left of the land face of the fort, they had a restricted option of swinging right around the western salient to enter the fort. Here they could attack the River Road sally port and "Bloody Gate" directly from across the sandy high marsh.

Though a slough certainly slowed their assault (the Confederates had previously removed the planking from a bridge over this water obstacle), the attack of Curtis's, Pennypacker's, and Dell's brigades successfully exploited the largest weakness of the fort. Once the Federals crossed the slough, they had the option of advancing up and over Shepherd's Battery, forcing the River Road gate, or bypassing these defenses altogether and slogging across the high marsh, bypassing the imposing earthworks. They chose to use all three approaches, tripling the strength of their push, and facing variable levels of rebel resistance.

This joint assault overwhelmed the poorly organized and distracted defenders, leading to the initial breach of the fort's parapet.

In his guide to the battles of Fort Fisher, Mark Moore suggested that the construction of the fort was not yet complete when it was attacked.[32] He suspected the River Road gate was an area intended to be strengthened. This critical weakness was first identified by William Lamb: "If the western salient, however, had terminated at the extreme edge of the river marsh—rather than on the road—it would have been difficult indeed for attacking infantry to pass around the end of the structure. And any access to the rear of the fort, despite a numerically superior enemy, would have been dearly won—especially if the parapet had been defended with sufficient manpower."[33] On Morris Island, no such options existed for the Fifty-Fourth Massachusetts. The salt marsh extended across almost the entire breadth of the island in front of the land face of Battery Wagner, and the high marsh zone on the back-barrier flank of the fort was all but nonexistent. At Fort Fisher the Federals could push across the mire at Federal Point in front of the left salient and move across the adjacent high marsh; Vincent's Creek behind Morris Island coursed directly adjacent to the sandy parapets, so any attacking Union regiment would need to have a boat to reach the fort. In short, at Fort Fisher the presence of high-marsh sediments made flanking a possibility; on Morris Island, flanking the parapet was an option that required swimming in either the Atlantic Ocean or Vincent's Creek.

PART VI

The Legacy of Sedimentary Geology and the Civil War

I will remark that the position indicated on the sketch is perfectly secure against the encroachment of the sea.
> —Captain William H. Chase, on selection
> of the site for Fort McRee in 1833

... It has been raining hard tonight, the wind is high, the waves of the Gulf lash and foam against the sides of our fort.
> —Captain James G. Bullard, stationed at Fort McRee, 1861

... Fort McRee and the somewhat anomalous structure to the westward are fast falling into the sea. They can no longer be considered as part of the defenses of the harbor.
> —Captain William E. Merrill, 1866

CHAPTER 18

Sedimentary Geology as a Tool for History

Sediments and Geoarchaeology

The science of geoarchaeology uses the tools and techniques of the geo-sciences to provide insights for historical interpretation. Techniques come from across all fields of the earth sciences: radiocarbon decay and dating, sedimentary particle-size assessment, micropaleontology, and near-surface geophysical analysis are but a few of the more common tools used in the field.

These geologically derived techniques can often be engaged to provide evidence for historians and archaeologists regarding controversial subjects from the past. For example, the exact location of the Battle of Thermopylae in 480 BCE has been muddied over the years by the complex interaction of inconsistent ancient historical accounts and later tectonic activity, sea-level fluctuations, erosion, and sediment deposition.[1] Two decades of geoarchaeological, stratigraphical, and paleo-environmental data collection, combined with extensive drilling, have provided a clearer picture of the location and width of the famous pass, in several cases confirming Herodotus's description of the local geomorphology.[2]

Geoarcheological studies from the American Civil War exist for battlefields from the Atlantic coast to Texas.[3] Two case studies illustrate the broad use of sediments and microfossils to solve Civil War archaeological mysteries and provide insights for historical interpretation. The first involves the Federal staging ground on Folly Island, South Carolina, for the 1863 assaults against Morris Island and Battery Wagner. Marsh microfossils and sediments were used to document the existence of a largely forgotten spring-tidal inlet that greatly complicated Gen. Gillmore's efforts at fortifying the northern end of the island.

The second case study's wreck site was found only seven and a half miles to the northeast of Folly: the sediment-filled hull of the world's first successful combat submarine, the *H. L. Hunley*. Sediments and microfossils from the interior of the vessel proved to be an incredibly important tool for understanding the truly remarkable preservation of the crewmembers' bodies—soft tissue intact after 140 years on the ocean bottom—after the submarine was recovered from the continental shelf in 2000, where it had been lying offshore of Charleston Harbor.

FORTIFYING FOLLY: SEDIMENTOLOGY AND LANDSCAPE RECONSTRUCTION

During the final two years of the Civil War, the Union navy and army worked in conjunction to reduce the permanent and temporary brick and sand fortifications surrounding Charleston Harbor. The costliest of these strongholds to capture was the sand fort Battery Wagner, located a mile from Fort Sumter on Morris Island, overlooking the southern entrance to the harbor.

To capture this island and battery the Federal forces first occupied the unguarded barrier island to the south, Folly Island. This first occupation occurred on February 8, 1863, when Maj. Gen. John G. Foster led a small reconnaissance force onto the southern end of the five-mile-long barrier. By mid-April thousands of Union infantry were occupying the southern and middle portion of the island. At this time the northern extension of Folly, which was within the artillery range of Confederate guns on the southern end of Morris Island, was left unoccupied.

Maj. Gen. Quincy Gillmore, the overall field commander for the operations against Morris, planned an amphibious assault from Folly to Morris Island via Lighthouse Inlet. To accomplish this, he would need to use the northern portion of Folly Island as a staging ground for his troops and as a gun platform for his covering artillery. To keep his planned maneuvers a surprise, he ordered this troop buildup and battery construction to be conducted in as stealthily a manner as possible. The geomorphology of the island, however, would make covert operations difficult.

The northern portion of Folly Island is also known as Little Folly or, delightfully, Rat Island.[4] It is isolated from the much larger southern

section of the barrier by a narrow strip of sand that was so low in elevation that it was often submerged by spring, high, or storm tides. Gillmore described this impediment to movement in the following manner: "extremely narrow, perfectly barren, and so low that the Spring
tides frequently sweep entirely over it. At the extreme north end, however, the sand ridges, formed by the gradual action of the wind and
tide, were, when our operations commenced, covered with a thick un-
dergrowth favorable for concealment and the masking of batteries." As
a result of this coastal geomorphology, the Federals on Folly could only
bring field pieces and supplies to Little Folly during low tide and at
night, to avoid the prying binoculars from across the inlet or unwanted
artillery attention. Negotiating this spring-tidal inlet during high tide
was difficult, and during spring tides or stormy weather, Little Folly was
essentially unreachable.

Nevertheless, by the end of April the northern edge of Little Folly
bristled with 3-inch ordnance rifles and dozens of heavy Parrott guns,
all concealed in the dunes and brush of the island vegetation. Of the
more than twenty thousand men to be stationed on Folly that spring
and summer, more than a quarter found themselves preparing for attack on Little Folly.

Despite the logistical significance of the spring-tidal inlet and the
elaborate, if concealed, earthworks on Little Folly Island, the location
of both historical sites had been lost through the decades. Coastal erosion and sea-level rise, and the accompanying island retreat, have certainly been responsible for the destruction of many of these landscape
features. Fort Green, for example, had been located on the extreme
northern end of Little Folly Island. The remnants of the fortification
were washed away by Hurricane Hugo in 1989.[5] Other sites, including
the inlet between Little and Big Folly and several gun batteries, still existed in the early 2000s, although their precise location on or below the
sediment surface was undocumented. This changed later that decade
after a series of geoarchaeological and stratigraphical studies based on
sediment distribution, and micropaleontological comparisons documented the location of the inlet and four long-forgotten and substantially eroded gun parapets.[6]

Five primary depositional sub-environments are found on Folly Island: low marsh, high marsh, beach/dune, washover fans, and inlets/
deltas. These sub-environments, or sedimentary facies, can be distinguished by their differing sediments and microfossil content. For ex-

ample, the low-marsh facies is composed almost entirely of clay and silt (mud) and almost always has tiny agglutinated foraminifers (microfossils) like *Miliammina fusca* and *Trochammina inflata*.[7] Beaches and dunes, in contrast, are composed entirely of medium and fine sands with a few abraded calcareous foraminifers mixed in with the sediment. Inlet sediments are a mix of sand and mud with abundant, pristine, calcareous foraminifers. The most interesting facies on or under Folly Island are those of the storm-generated washover fans. These sandy sediments contain open-marine foraminifers, usually eroded from the continental shelf during hurricanes. These sands and microfossils get carried into the back-barrier marshes during large storms, where they spread across the marsh (thus, a fan). Later the sandy washover layers are reburied by the encroaching marsh, burying them in mud as sea level and the marsh surface continue to rise. The marsh strata behind Folly Island contains more than a dozen of these storm deposits, all interbedded between layers of marsh mud.[8]

These five modern facies were used to relocate the strategically important spring-tidal inlet that divided Folly Island during the Civil War. Fifteen-foot-long auger cores were taken in areas where the inlet was suspected of having been located to see if it could be detected in the subsurface using grain size and microfossil analyses (see figure 18.1). All five depositional environments were recognized in the cores, but only one location, near the modern cul-de-sac that ends Ashley Avenue, contained an array of inlet-indicative sediments and fossils. Additional coring at this location outlined the lateral extend of the small inlet, which is today buried under 6 feet of marsh mud and migrating dune sand.

Most historians had placed the location of the inlet in the region of the island known as "the washout."[9] This fragile portion of Folly Island has been blown open multiple times in the past, including during Hurricane Hugo. However, this is not the location of the inlet described by Quincy Gillmore. The Civil War spring-tidal inlet was instead located almost a mile farther along the island to the north—and much closer to the Confederate artillery across the considerably larger Lighthouse Inlet.

Sea-level rise, island migration, and shoreline armoring (groins) had acted in conjunction to hide this landscape feature; geoarchaeology, in the form of sediment and microfossil analyses, relocated it, clarifying at least one minor—but tactically and logistically significant—mystery in the historical record of the Charleston Campaign.[10]

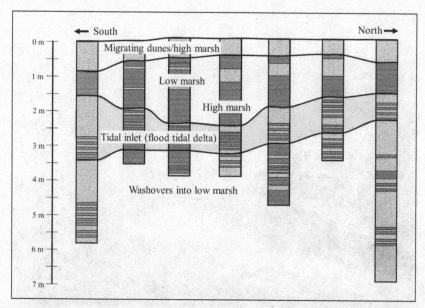

FIG. 18.1. This cross section represents the strata in the immediate back-barrier of Folly Island where a strategically important spring tidal inlet was located. The facies transition from washovers into marsh muds (prior to 1860's) into the inlet (mid-19th Century), before the marsh muds and dune field buried the inlet sediment (1900–present).

A second contribution by sedimentary geology to the forgotten Civil War history of Folly Island did not require the analysis of microfossils or sediment grain size. Instead, sedimentary structures, and specifically cross-bedding, were used to differentiate natural sand dunes and sandy relict Civil War gun parapets (see figure 18.2).[11]

Cross-bedding forms during the formation of ripples or really big ripples (dunes) when the original bedding plane is tilted, producing sets of parallel sloping layers of sediment. The extensive dune field on Folly Island has mounds of sand that, without exception, exhibit cross-bedding. However, four round/oval piles of sand adjacent to Lighthouse Inlet lack any form of sedimentary structures. While these automobile-size piles of sand may appear to be small sand dunes, their absence of cross-bedding indicates they were not created by wind; instead, these were made by shovels. As Federal soldiers and sappers reallocated the dune sands for parapets and artillery lunettes, they destroyed the original cross-bedding, leaving only a massive pile of amorphous sand behind.

Gouge-auger coring of the sand mounds quickly identified four

FIG. 18.2. Lighthouse Inlet separates Folly Island (left center) from Morris Island (distance). Sedimentary geology documented the position of four gun lunettes (1), while an ebb delta marks the former position of the southern edge of Morris Island (2). Battery Wagner's position (3) is now in the Atlantic.

distinct anthropogenic "dunes," all located at the site of a Federal 30-pounder Parrott gun battery.[12] The size and relative orientation of these mounds relative to the Confederate positions across the inlet also suggests they are of Union construction, instead of aeolian.

Coastal erosion continues to plague Folly Island, and storms continue to reveal long-buried artifacts and destroy landmarks. Cannonballs were uncovered by Folly's shifting shoreline after both Hurricane Matthew in 2016 and again after Hurricane Dorian passed by in 2019.[13] Storm waves and winds have also removed what was left of the Civil War lunettes, all identified in the early 2010s and, regrettably, destroyed within a few years.

SEDIMENTS AND SUBMARINES

When the Confederate *H. L. Hunley* sank the sloop of war USS *Housatonic* outside Charleston Harbor on the evening of February 17, 1864, it became the first submarine in history to sink an enemy vessel. Success was stunningly brief, however, as the boat and all hands disappeared the same evening.

When the submarine was discovered in 1995 and recovered in 2000, it was clear the archaeologists, geologists, and historians had an exceptionally large, valuable, and fragile artifact with an amazing story locked within the sediments filling the hull. The condition of the remains of the crew, for example, was astounding: soft tissue and intracranial material were present even after sitting on the continental shelf for well over a hundred years.

The *Hunley*'s wreck and recovery site is located in 25 feet of water approximately four and a half miles east-southeast of the entrance to Charleston Harbor. The same set of jetties that led to the rapid erosion of Morris Island to the south of the harbor also helped bury the submarine, channeling sand that was destined for Morris, and to a lesser extent Folly Island, farther offshore onto the shelf. The *Hunley* came to rest on a heterogeneous mixture of reworked Neogene, Pleistocene, and Holocene sediments—the same mixture of sediments found in the washover fans of Folly Island.[14] Modern shallow-marine sediments (fine sand, clay mud, and shell material) cover these older sediments and eventually buried the boat. Calcareous foraminifers are especially abundant in these sediments, with hundreds of specimens found in every cubic inch of sand and mud (figure 6.1).

The sediment-infilling history is a complex stratigraphic puzzle that was first pieced together by M. Scott Harris of the College of Charleston.[15] He used the variations in sediment type, sedimentary structures, and macrofossils to construct a four-step process for sediment deposition within the hull. The infilling history of the *Hunley* is complicated by three breaches that occurred to the hull: one round 5-inch-diameter hole in the forward conning tower over the captain's position that may represent combat damage, and two much larger openings on the bow and stern that were the result of sand scouring by water currents while the submarine rested on the high-energy seafloor.[16]

The first layer of sediment deposited within the submarine almost certainly entered the hull through the fracture in the conning tower. This cone-shaped blanket of interlayered mud and silt draped across the captain's position at the front of the crew compartment.

The second layer of sediment to form within the submarine was the most important with regard to the exceptional taphonomic condition of the crew.[17] This layer, known by those working on the submarine as the "toxic mud," completely buried the bodies of the vessel's commander, Lieut. George Dixon, and the seven crew members. This layer

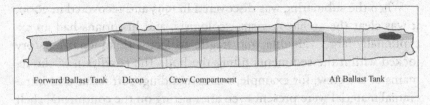

FIG. 18.3. Cross-section through the *H.L. Hunley* showing the density of foraminifers (darker shading equates to more microfossils per cubic inch of sediment). Black ovals represent the three breaches to the hull. Note the foraminifers are most abundant in the "healthy sand" and largely absent from the lower "toxic mud" that encapsulated the crew members' bodies. "Dixon" indicates the position of G.E. Dixon, the commander of the submarine on its final mission.

was bereft of macrofossils, suggesting a harsh, low-oxygen, scavenger-free depositional setting.

The third layer of sediment deposited within the submarine was similar with respect to grain size and macrofossil content to the sediments surrounding the submarine at the wreck site. Harris demonstrated that this layer was created after the two large hull breaches had formed.[18] An internal cross flow eroded the top of the toxic mud and deposited coarser sand and abundant mollusks, sponges, and echinoderms. Had the crew members' bodies been encapsulated in this layer of sediment, instead of the now buried mud, it is likely archaeologists would have found only scattered and scavenged bones and teeth.

The final layer of sediment to accumulate within the hull was created after the healthy sand filled the submarine to the point that the fore and aft breaches were blocked and water currents within the hull ceased entirely. This layer is primarily mud and similar in many sedimentological aspects to the second layer, sans corpses.

Nearly twenty-five thousand microfossils were collected from the sediments within the *Hunley* to test Harris's infilling hypothesis and provide insights into the exceptional preservation of the crew (see figure 18.3).[19] The dissimilarity in microfossil assemblages between the toxic mud and overlying healthy sand layers is striking. The toxic mud, unsurprisingly, has few foraminifers of only a few species. One taxa that is found in quantity in this strata is *Bolivina* spp., a genera known to prefer low-oxygen environments.[20]

The healthy sand, in contrast, has a microfossil assemblage nearly identical to that of the surrounding continental shelf: abundant foraminifers with great diversity and copious echinoid spines, sponge spicules, and diatoms. In all manners the microfossils corroborate the sediment-infilling model, documenting an anaerobic setting for the initial burial of the crew followed by a more significant breaching of the hull with shelf sediments (relatively) rapidly filling the hull.

DIATOMS AND MODES OF DEATH

Diatoms are marvelous microscopic aquatic plants that create ornate siliceous tests called frustules (see figure 18.4). Of all the varieties of microfossils, diatoms are found in the most environments, including open ocean, marginal-marine (marshes/lagoons), rivers, and lakes. They are also incredibly abundant.

The tiny size of the diatom frustule and cosmopolitan nature of the critter makes them a potentially valuable forensic tool.[21] In northern European criminal courtrooms, for example, diatoms have been used to differentiate the mode of death for a corpse between drowning and suffocation. In North America the forensic use of diatoms is more limited.[22]

The methodology, analysis, and reasoning behind the use of diatoms as an indicator of cause of death follows this reasoning: When a person drowns, their final involuntary action is to rapidly inhale, whether they are submerged or not. When water—and presumably diatoms—are inhaled, the unfortunate person no longer has oxygen to breathe because the lungs are filled with fluid. Nonetheless, the heart continues to beat, circulating blood throughout the body. Diatoms are small enough to pass through the lung membrane and enter the circulatory system, thus being distributed to the organs and other soft tissues during the final moments of life.

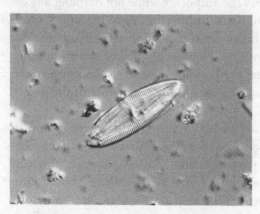

FIG. 18.4. This siliceous pennate diatom frustrule was collected from the *H.L. Hunley* wreck site. The microfossil is 40 micrometers long.

When a person asphyxiates or is suffocated, in contrast, and the body is disposed of in a body of water, the heart will have long since stopped beating before the lungs eventually fill with water. As a result, there is no viable mechanism to distribute diatoms from the lungs to the other internal organs or tissues. Diatoms may be in the saturated lungs, but without cardiovascular circulation there is no way for them to pass through the lung membrane into other parts of the body.

In summary, then, when a body is found in a diatom-rich pond or lagoon, the presence of diatoms in the organs (commonly the liver) points toward inhalation of diatoms and water while drowning. The absence of diatoms suggests possible suffocation, but the lack of evidence is not as definitive a diagnostic tool for determining the mode of death.

The crew members of the *H. L. Hunley* are in remarkable taphonomic condition, especially considering their age and the harsh nature of the wreck site's depositional environment (fast currents, abundant scavengers). They are, however, far from the putrefied corpses studied in European forensic-microfossil legal cases. *Skeletonized* would be a more accurate term for the *Hunley*'s crew. Nevertheless, soft tissue does exist for several of the crew, including adipocere and intracranial material (deteriorated brain matter).[23] The adipocere from one member of the crew was recovered from the interior of a femur and tested for the presence of diatoms using a multi-proxy approach, all with hopes of providing insights into whether the crew drowned or suffocated.[24]

The key challenge to identifying the presence of diatoms in the corpses is the nature and composition of the waxy adipocere; it must be separated from the siliceous microfossils for proper analysis. Siliceous fossils are relatively chemically stable, so a variety of methods were used to isolate or destroy the adipocere while not harming any diatoms that might be present. These approaches included simple enzymatic digestion, the methodology most commonly used in criminal investigations across the Atlantic, and a more aggressive approach using diluted sulfuric acid. The later methodology was needed because of the more unusual skeletonized, rather than putrefied, nature of the potential diatom source material.

After a five-year-long study ended in 2019, diatoms were found in the surface and bottom waters from the *Hunley* wreck site and the sediments from all strata within the submarine. Diatoms were not, however, found in the adipocere of the femur. If the microfossils had been found in the adipocere, the presence of the microfossil might have provided evidence, in a highly limited manner, that the crew drowned. The lack of diatoms in the adipocere, however, does not necessarily support suffocation as the cause of death. The lack of diatoms in the adipocere only indicates that the men either (1) did not inhale water, or (2) the water did not contain diatoms, or (3) the diatoms did not escape the lungs. In other words, the presence of diatoms in the adipocere might have been a qualified indicator of mode of death (vulnerable to contamination); the absence of diatoms indicates very little. With those caveats extended, then, scientists are left pondering what other non–sediment-based method might be used to differentiate drowning from suffocation using only skeletonized material from cadavers from a submarine that sank almost 150 years ago.

CHAPTER 19

The Fate of the Fortifications

Evolution of Battleground Landscapes

Sand and sediments can provide insights into historical mysteries, but sedimentary processes can also destroy history. The 150 years since the fighting ended do not register on the geologic time scale, yet significant changes have occurred to the battle terrain during this time. The most disheartening alteration to the landscape is (was) also the most preventable: anthropomorphic alteration.[1] Highways, shopping centers, and housing developments now cover much of the land where significant battles occurred, forever changing the true lay of the land. As noted historian Gary Gallagher stated, "No historical landscape is immutable. Natural and built features change regularly, sometimes for the better and sometimes not."[2]

Nature has certainly claimed multiple important battle sites and fortifications as well, with fluvial and coastal erosion continuing as the primary culprit. Rivers and ocean waves truly define the use of the term "temporary" in temporary fortifications. In a broad sense, the destructive processes that remove earthworks and fortifications from the landscape can be subdivided into two groups: rivers and coastal erosion represent *dynamic destruction*, capable of obliterating a site in a few years. These largely nonmitigatable forces destroyed Battery Wagner and Island Number 10 within a decade or two after the fighting stopped in 1865. A second mode of destruction is *slow and steady*: erosion over many decades by wind, precipitation, and runoff.[3] This is the largest preservation challenge to the National Park Service, which has spent a half century investigating the best way to slow the effects of this type of weathering on surviving archaeological sites.

At least some degree of alteration and erosion has occurred to all battlegrounds, even those that have been designated for protection and conservation over the years. The Vicksburg National Military Park is an interesting study in the battle between erosion and conservation efforts by the federal government. The military park was established in 1899, and the government has owned the 1,800 acres and 20+ miles of trenches and earthworks ever since, giving the National Park Service control of preservation efforts in 1933.

The Confederate earthworks that were so critical to the defense of the city had been neglected during the first year of their existence. The silty loess that underlies and composes the fortifications is especially prone to erosion from runoff after it has been disturbed by digging, and the mass wasting that began in 1862 continues to plague the battlefield today.[4] As recently as 2017 the National Park Service closed Confederate Avenue and North Union Avenue after heavy storm runoff from Hurricane Harvey eroded an 8-foot-deep gully under the roads.

Efforts to minimize erosion at Vicksburg National Military Park began in earnest in the 1930s. The Civilian Conservation Corps planted thousands of trees across the battle landscape, hoping the deep roots would stabilize the soil and prevent soil creep, slumping, and landslides.[5] These efforts exposed the conundrum that still exists for preservationists on the modern battlefield: many efforts to minimize erosion and protect earthworks also make it more difficult to visualize and appreciate the historical landscape. Multiple recent studies demonstrate the preservation potential of forest cover when compared with the exposed, grass-covered fieldworks seen on many battlefields today, but earthworks overgrown with trees and shrubs are difficult to interpret from a historical or tactical perspective, and battle sight lines are compromised by tree cover.[6]

The National Park Service has determined that two categories of earthworks exist on Civil War battlegrounds: those that are minimally managed in forests and those that must be actively managed in open terrain.[7] These latter earthworks are typically covered with manicured lawn grasses. For these "remnant" earthworks, the choice of vegetative cover is largely controlled by soil type.

Damage and erosion of the forest and grass-covered earthworks comes from a variety of different natural and anthropogenic sources. In forests,

FIG. 19.1. Erosion of Confederate earthworks at Grand Gulf National Military Park. In the foreground a gully has formed across the parapet guarding the lower portion of Fort Wade. In the background large trees uprooted by tornadoes in the spring of 2019 have damaged trenches and parapets.

very large trees may uproot, damaging significant portions of trench lines or parapets as the falling tree's broad network of roots creates its own crater. Earthworks in wet soil or covered with shallow-rooting trees are especially vulnerable to this form of destruction (see figure 19.1).[8]

The preservation of Civil War trenches is directly related to sedimentary geology and the strength of the soil where the trench was excavated. Contrast, for example, the remnants of trenches cut by the armies into the Coastal Plain sediments at Fredericksburg and Petersburg more than 150 years ago with those dug into the Normandy coastline by the German army during the second half of World War II. The cohesiveness of the clay in the Virginia soils helped preserve the Civil War trenches, preventing complete collapse and flattening, while the nearly pure sand from the bluffs above the D-Day beaches quickly collapsed because of a lack of strength in the soil (see figure 19.2).

Nevertheless, the long-term durability of earthworks provided by the clay in the soils also had one negative influence on earthwork survival: clay-rich Coastal Plain soil is better for agriculture than coastal sand. Many earthworks were lost to the plows of farmers.

FIG. 19.2. The shallow zigzag depression running down the left-center of this photograph toward Omaha Beach is all that remains of a 5½ foot deep German communications trench at WN 62 ("Wiederstandnest" or point-of-resistance, number 62). The monument at the bottom of the slope honors the 5th U.S. Engineer Special Brigade.

One final erosional influence on earthwork preservation is perhaps the saddest. Reckless digging from relic hunters destroys the delicate soil stratigraphy of the field fortifications, creating completely unnecessary cratering to the large and fragile earthen artifacts. The scars from this often illegal activity mar battlefields like Cold Harbor or Petersburg for decades after the theft.[9]

Erosion and Preservation

The right combination of friction (sand) and cohesion (clay) in a sediment can provide a very strong material for use in field fortifications. As demonstrated on Morris Island, sand content is more important than clay content for defeating shellfire; unfortunately, the mature quartz sand on Morris Island that was demonstrated to be so strong for protection from gunfire was also exceedingly vulnerable to erosion.

Erosion is the transport of weathered material, and sand is, by definition, weathered material. It is easily moved by water (rainfall, runoff, rivers) or wind, necessitating either the planting of marsh grass or sea oats on sandy earthworks or covering the parapets in squares of sod for stability.

Coastlines are incredibly dynamic sedimentary environments, and erosion occurs at many scales and from many causes. Local sea-level rise will slowly drown a coastline over a decade while a hurricane might reshape a region overnight. A barrier island's natural response to sea-level rise and storms is to retreat toward the mainland.[10] A process called overwash spreads sand from the beach and dunes across the back-barrier marshes and lagoons, establishing a platform for landward movement (i.e., retreat) by the island. Americans spend hundreds of millions of dollars every year trying to slow or stop this natural transgressive process through armoring of the shoreline with sandbags or concrete or beach replenishment projects.[11]

Coastlines underlain by particularly hard rock (igneous and metamorphic) or sedimentary rock, like sandstones, conglomerates, or shales, are the most resistant to shoreline erosion from waves. Coastlines that are composed of semi-lithified sedimentary rocks (partially or weakly cemented clastic particles) are less resistant and have higher rates of retreat.[12] Coastlines that are exclusively composed of sediments (usually medium sand) or soft rocks like chalk are the most vulnerable to erosion and retreat.

Most of the coastal fortification along the Atlantic and Gulf coasts are underlain by unconsolidated sediments or a combination of unconsolidated sediments and partially lithified sedimentary rocks. Around Charleston, for example, the modern barrier islands are composed of sand (beaches, dunes) and mud (back-barrier marshes). In the nearshore area, however, outcrops of weakly cemented Oligocene and Miocene clay-rich sands make up formations of rock that are between 5 and 40 million years old. This heterogeneity of sediments makes predicting erosion highly complicated.[13] The direction of incoming waves, the supply of sediment, the rate of local sea-level rise, and the frequency of storms all play a role in the rapidity of shoreline erosion and retreat.

Along the modern Atlantic and Gulf coasts the rate of erosion has increased in the last hundred years, and it will likely continue to do so in the future. Decadal rates of sea-level rise were measured in inches during the nineteenth century and feet during the twentieth.[14] This increasing rate, combined with a reduced sand supply for beaches from now-dammed rivers, has led to the loss of multiple important Civil War sites.

In the 150 years after the fighting stopped, many Civil War coastal fortifications were either completely lost, such as Battery Wagner, or partially destroyed, such as Fort Fisher, because of shoreline retreat.[15] While the reasons for increased erosion for each site are complicated, the primary factor or factors responsible for loss can often be identified, and usually the most important cause is anthropogenic. Battery Wagner, for example, was doomed from the moment the U.S. Army Corps of Engineers (USACE) finished building the Charleston Harbor jetties around the turn of the twentieth century.

Sea-level rise is the one consistent element affecting all coastal fortifications from the southeastern United States. Sea level has risen almost a foot since the Civil War. As a result of the flat topography of the Coastal Plain along the Atlantic coast, the relationship between the rate of sea-level rise and shoreline retreat can be as high as approximately 1:1000.[16] In other words, using this simplified rule, if an island experiences one foot of local sea-level rise, the beach and island would be expected to migrate 1,000 feet toward the mainland. The back-barrier marsh and transition zone between marsh, lagoon, and mainland would all shift landward across the flat landscape as well.

Without expensive and temporary artificial beach replenishment or shoreline armoring (seawalls, breakwaters, groins), this rate of retreat would mean the loss of nearly all coastal sand forts.[17] Most of the fortifications were, after all, constructed as closely as possible to the high-tide line to maximize the effectiveness of ricocheting fire across the water. Hurricanes, tropical storms, and nor'easters only exacerbate the shoreline erosion.

Permanent Becomes Temporary

Temporary fortifications were built during wartime and lost much of their strategic importance after the fighting moved elsewhere. Permanent fortifications were built during peacetime and were situated and constructed to have strategic value for decades. During the Civil War, the vulnerability of each type of fortification to enemy fire was demonstrated to be related to the choice of construction material, and not necessarily the fort's designated life span.

Forts Pulaski and Macon fell quickly in combat but survive today. The military engineers who selected their permanent sites, including Robert E. Lee, chose well: the center of broad, relatively stable barrier or delta islands far from rapidly migrating inlets.[18] This site selection, which represented a careful deliberation between artillery range and foundation stability, accounts for much of the survivability.[19]

Reconstruction of Fort Sumter began in 1870 under the direction of, paradoxically, Quincy Gillmore.[20] Only one tier of casemates remains today, and much of the interior of the fort is occupied by a massive concrete gun emplacement for two 12-inch rifles that were added in 1899.[21]

These brick forts took decades to build and will most likely survive for another century. The temporary fortifications were constructed in months and lasted, in almost every case, only decades. The primary culprit in this destruction was erosion from beach drift and storms. Both Battery Wagner and Fort Fisher were sited as close as possible to sea level, primarily so they could make maximum use of their heavy guns on distant enemy ships. This low elevation necessitated their construction very near the high-tide line, with predictable geologic consequences. Fort Fisher had a palisade that extended into the surf, and Battery Wagner had a tidal moat that connected directly to the Atlantic. Ocean water and salt marshes limit the avenues of attack for the enemy, but both are also, hazardously, found at sea level.

As sea level has risen over the last 160 years, the ocean has transgressed closer to, or over, the site of the earthworks. Other nineteenth-century historic structures like the Cape Hatteras Lighthouse or the Belleview Biltmore Hotel (near Tampa, Florida) can be relocated, at great effort and expense, when the shoreline retreats. Sand forts like Fort Fisher cannot.

Fort Fisher, the famous protector of the Cape Fear River and port of Wilmington, North Carolina, has been more than half destroyed by coastal erosion. The shoreline here appears to follow the 1:1000 rule as the shoreline has retreated nearly 600 feet in the hundred years after the war, destroying the entire sea face of the South's largest fortification. Attitudes toward preservation of the eroding fort have changed through the decades. During World War II a significant portion of Fort Fisher's parapets were flattened for an aircraft runway. Ten years later, hard armoring was added in a desperate attempt to stabilize the shoreline and save the remaining fragments of the fort. The state and USACE constructed hardened revetments of increasing size during the 1950s, 1970s, and 1990s (see figure 19.3).[22]

In 1865 the Federal navy attempted to destroy the fort, starting with a bombardment of the land face and working south. They made only su-

FIG. 19.3. Battling the beach drift. This
revetment at Fort Fisher protects what
is left of the landface of the fort. In 1865
the fort's landface parapet would have
extended across this picture, before
joining the seaface to the right.

perficial damage to the sandy parapets. The Atlantic Ocean took the opposite geographical approach to destruction, working slowly from the
south to the north, eroding the sea-face parapets first. The Mound Battery was the first to be lost, eroded away before the turn of the twentieth century. Erosion continued into the 1990s until the last third of the
fort, the western land face, was temporarily protected by a granite riprap revetment. The armoring was deemed necessary by the USACE because of the rapid rate of shoreline retreat: almost 10 feet per year in the
fort's sector of the beach. Coastal geologist Orrin Pilkey points out that
engineering structures such as this revetment exacerbate erosion problems on nearby beaches: "The wall was built to protect Civil War earthworks, but several decades from now erosion rates will greatly increase
on the adjacent beaches in both directions because of the wall. Civil
War forts, lighthouses, and many other structures have been allowed to
fall into the sea elsewhere. As precious as Fort Fisher may be, its preservation may prove to be a pyrrhic victory."[23]

Battery Wagner has a similar sad story, but one with an ironic twist.
Quincy Gillmore wasted time, ammunition, and lives attempting to destroy the fort during the third year of the Civil War. Twenty years later
he would quickly and quietly destroy it using granite.

Gillmore's engineering specialization involved utilizing concrete
along the shoreline, and after completing his service during the war
he went on to direct several harbor improvement projects. In the 1880s
the federal government tasked him with improving the seaports along
the southeastern Atlantic coastline, many of which he had earlier led
campaigns to capture. When Gillmore returned to Charleston, his pri-

mary efforts were poured, so to speak, into improving the navigation channels into the harbor with concrete and boulders. To stabilize the ebb delta and shifting shoals and channels, the USACE constructed two massive jetties, extending from Morris and Sullivan's Islands offshore for three miles. These piles of boulders would allow the channel between the rocks to be dredged, stabilizing the once-dynamic navigation channel into the port.[24] The southernmost of these jetties was constructed in 1880 and intersected with Morris Island at Cummings Point.

Morris Island, already a site of rapid erosion, was now starved of sediments carried from the north by beach drift and the longshore current. Sediments carried by the rivers into and through Charleston Harbor were also funneled offshore by the new jetties and the strong ebb currents. By the time the jetty project was complete in the mid-1890s, Morris Island had retreated to the point where little of the sacred ground the men had fought over had survived. Sand from Morris was carried south to Folly Island and Edisto Beach but was not replaced by drift from the north. The famous Morris Island Lighthouse, built in 1876 in the dune field at the center of the island, was on the beach by the start of World War I and in the water by World War II; today, it sits more than 1,000 feet offshore. Within a decade of the completion of the jetties, all coastal sand fortifications on Morris were gone, and Batteries Wagner and Gregg's sediments were scattered into Lighthouse Inlet and along Folly Beach to the south. Beach drift and shoreline retreat, not heavy artillery, would remove any trace of the fort's existence.

Batteries Wagner and Gregg on Morris Island, South Carolina, were completely destroyed through a combination of sea-level rise and a nearly complete shutdown of the sand supplied by natural beach drift. Wind-driven waves almost never strike the beach at a completely perpendicular approach (with the crest parallel to the beach). Instead, as they enter the surf zone at an angle they push water up the beach slope following an arched path—the swash up and onto the beach is at an angle, but as the wave slows the return backwash moves straight back down the beach, perpendicular to the trend of the shoreline. The resulting current will create a shallow water flow offshore called the littoral current that parallels the beach in the direction of the incoming wind and waves. On the beach, sand particles are transported by the breaking waves and swash in this same direction, meaning the entire portion of the beach in contact with the water in the surf zone is slowly migrating down-drift along the shoreline. If the "up-drift" supply of sand is restricted for any reason, the result will be erosion and shoreline retreat of the down-drift portion of the beach.[25]

On the barrier islands surrounding Charleston Harbor the direction of beach drift is generally toward the south. North of the harbor, sand from Sullivan's Island passes Fort Moultrie before accumulating in the flood and ebb deltas at the harbor entrance. The sediment is then transported by waves and currents southward onto Morris and Folly Islands. In the nineteenth century the shifting sands in the Charleston Harbor deltas constantly modified navigation channels, resulting in numerous ships running aground.[26] To rectify this problem the USACE decided to stabilize the harbor inlet with a pair of massive jetties. After their construction was completed in 1896, the normal beach drift was highly diminished, much to the detriment of the islands to the south, including Morris. Folly Island, the next island to the south, also suffered from a lack of sediment coming from Morris, and today it is the most heavily armored island in South Carolina.[27]

Beach drift carried sand from Sullivan's Island along the jetties and farther offshore into deeper water. There the sediment was no longer a threat to shipping, but it was also too deep to be moved by normal wave activity.[28] Additionally, sand carried by the Cooper and Ashley Rivers was also funneled further offshore by the jetties, bypassing the beachfront of the batteries.[29] The resulting shutdown of beach drift by the new jetties led to the loss of what was left of the fortifications within a few decades and a thinning of the island.

Oblique aerial photographs of Cummings Point on the northern end of Morris Island show that the location of Battery Wagner is today offshore (see figure 19.4). A shoreline engineering structure known as a groin currently marks the former location of the fort. Ironically, it was built to slow the retreat of the island by capturing beach drift—the same phenomenon that caused the loss of the fortification when the jetties were built.

Beach drift can completely alter a historical landscape, rendering the interpretations of the relative locations of Civil War fortifications and ships confusing. For example, Forts Pickens and McRee were constructed on opposite sides of the inlet that allowed access to Pensacola Harbor and an important naval yard. Both brick fortifications were constructed as close as possible to the ends of their respective barrier is-

FIG. 19.4. Cummings Point, Morris Island, South Carolina and the location of Battery Wagner (in the shallow water near the groin and small boat). Charleston Harbor and Fort Sumter are in the background. Note the slight accumulation of sand to the north of the groin—this is an indicator of the direction of natural beach drift to the south, away from the harbor entrance.

lands to offer cross fire over the inlet. However, beach drift and storms created problems for both forts even before the works could see combat. Fort McRee was constantly threated by sand drifting away from the beach to its front; by 1866 the fortress was falling into the Gulf.[30] Fort Pickens had exactly the opposite problem; beach drift carried sand along the shoreface, extending the spit of sand between the fort and the inlet.[31] Eventually, the guns of the fort were more than a mile away from the inlet but still directed toward the now-submerged position of Fort McRee.

One solution for saving historical shoreline structures is relocation. The preservation of Cape Hatteras Lighthouse is a fine example of this expensive but effective strategy. The view from the top of the lighthouse demonstrates the vulnerability of the original location to shoreline erosion and the distance of transport for the relocated structure (see figure 19.5).

Relocation isn't possible with massive masonry structures or fortifications composed of unconsolidated sediments. The past and future of these structures can be envisioned by comparing the fate of the Cape

FIG. 19.5. View from the top of Cape Hatteras Lighthouse towards the original site (on the beachface). Recently, even the monument marking the original location had to be moved as it was threatened by erosion (crescent of curved blocks, center).

Hatteras Lighthouse to that of the lighthouse built on the south end of Morris Island, 1,300 feet from the water's edge. Originally constructed in 1876 in the dunes on the highest and most protected portion of Morris Island, the lighthouse today stands nearly a thousand feet offshore (see figure 19.6). The sand that would have (partially) stabilized the beach surrounding the lighthouse sits instead offshore of Charleston Harbor. Without armoring, and a concomitant loss of the beach, or continuous replenishment, a similar fate awaits barrier island fortifications like Fort Macon, North Carolina, Fort Morgan and Fort Gaines near Mobile, Alabama, and Fort Moultrie, on Sullivan's Island outside of Charleston, South Carolina.

FIG. 19.6. Morris Island Lighthouse under repair in Lighthouse Inlet. The retreating remains of Morris Island are in the background. Had a similar oblique aerial photograph been taken when the lighthouse was constructed, no water would appear in the picture.

At the Mercy of the Meanders

The forts and earthworks from the Mississippi River Valley vary greatly in their state of preservation. For those that are completely gone, Island Number 10, Fort Henry, and Fort Hindman, the primary agent of destruction was fluvial erosion or flooding. Rivers are highly dynamic environments, and the meandering Mississippi and Arkansas are constantly cutting new destructive erosional paths. These Confederate fortifications were constructed to overlook and control a river, and thus they were built directly adjacent to the water, making them especially vulnerable to cutbank erosion.

Island Number 10 was almost completely eroded away by the mid-1870s. In his novel *Life on the Mississippi* (1874), Mark Twain described what was left of the island:

> I found the river greatly changed at Island No. 10. The island which I remembered was some three miles long and a quarter of a mile wide, heavily timbered, and lay near the Kentucky shore—within two hundred yards of it, I should say. Now, however, one had to hunt for it with

a spy glass. Nothing was left of it but an insignificant little tuft, and this was no longer near the Kentucky shore; it was clear over against the opposite shore, a mile away. In war time the island had been an important place, for it commanded the situation; and, being heavily fortified, there was no getting by it. It lay between the upper and lower divisions of the Union forces, and kept them separate until a junction was finally effected across the Missouri neck of land; but the island itself joined to that neck now, the wide river is without obstruction.

Remnants of the poorly sited Fort Henry lasted until the Tennessee River was deliberately dammed in the 1930s by the Tennessee Valley Authority, and today what is left of the earthworks is below Kentucky Lake. Fort Hindman is also completely gone, the result of the meandering of the Arkansas River.

Other fortifications survive, but in a much altered state or often, unfortunately, in poor condition. After the Civil War the Arkansas River changed course, destroying much of what remained of Fort Pillow.[32] Only a portion of the defenses of Port Hudson are visible today, but not because of the river; urbanization has tragically overrun the earthworks, encroaching on what remains of the battleground.

Three Civil War sites stand apart regarding their preservation, and all are more than 400 acres in size. Fort Donelson National Military Park was established in 1928, protecting the riverfront batteries and 1,000 surrounding acres. Grand Gulf Military Monument Park, which has a rather redundant-sounding name, protects about half as much land as the Fort Fortress Donelson location site. The community of Grand Gulf is gone, however, as the town became a ghost of its former self around the turn of the twentieth century because of a shifting of the Mississippi River's course.[33]

Vicksburg also suffered a decline in river commerce with the meandering of the river. Here the earthworks and battleground sprawl across two sacred square miles, averaging nearly a monument or marker per acre. Not all of the early explanatory plaques survive today, however. Interestingly, many were sacrificed for a different war: nearly 150 of the largest commemorative iron markers were melted down in 1942 at the bequest of the War Department, only to be reborn as arms and equipment for the Second World War. Fortunately, a vast array of markers and artillery pieces still exist on the battleground today, identifying the multitude of batteries and troop positions for visitors.

The variable preservation of these river fortifications should not be surprising; they were, after all, temporary fortifications constructed of sediment. While many were saved by the federal government from the bulldozer of development starting in 1899, fluvial erosion was a natural force too great for the preservationists to prevent. The same can, unfortunately, also be said for another set of sedimentary fortifications, the battle-proven sandcastles from the coast. Here, as in the Mississippi River Valley, the primary agent of destruction was moving water, and today only a hint of the fortifications' strategic value remains visible on the beaches and barrier islands.

CHAPTER 20

Lessons Learned for Future Conflict

Sediments:
Grain Size Determines the Force Multiplier

One needs to look no further than the barrier islands of the Carolinas to see the clearest example of sand as a defensive force multiplier. The Confederacy was able to repulse repeated infantry incursions and naval and land-based artillery assaults with undersized garrisons, all while inflicting terrific losses on Federal infantry on the beaches before Battery Wagner and Fort Fisher.

The geomorphology of these islands also restricted the size of the Union attacking force, diminishing the strategic value of their numerical superiority. Infantry cannot attack across marsh mud or in ocean water, so the direction of an impending assault is entirely predictable: artillery can be oriented and aligned to reflect this path of attack, allowing a concentration of whatever firepower existed within a fortification with little risk of needing the guns along another sector of the sand works.

The silt and mud of the Mississippi River Valley offered clear benefits for both offensive and defensive strategies, and the tactical value for or against a maneuver can again be discriminated based on grain size. The loess at Vicksburg provided the Confederacy with river-cut bluffs for strong, elevated artillery positions and a material that was easy to modify into elaborate redans, lunettes, and trenches. These defensive benefits were somewhat balanced by the ease with which Grant's men could construct approach saps and mines, as well as elaborate encompassing (and isolating) trenches of their own. These easily constructed and formidable earthworks were never directly assaulted by the rebels, and their spatio-lateral totality essentially quarantined the city and Confederate garrison, assuring no food or supplies would penetrate the lines. As environmental historian Lisa Brady summarized, "Vicksburg's steep bluffs and eroded loess hills initially served Pemberton's army as a sort of ready-made fortress, handily keeping the Union forces out. In the end, however, those same features became prison walls, trapping the Confederates inside."[1]

Mud, just like the silt of Mississippi loess, also offers both defensive and offensive military advantages—or at least the distribution of the fine sediment does. The same soft mud and geomorphology that made Grant's maneuvers down the western side of the Mississippi slow and tedious also allowed his engineers to attempt new and innovative approaches to bring combatants to the proximity of the Confederates in the Gibraltar of the South.[2] Eventually, it was the combined approach on river levees, with additional troops and supplies carried rapidly downriver past the heights of Vicksburg, that allowed Grant's army to cross onto the loess and conduct his brilliant operations on the Coastal Plain in central Mississippi.

In contrast, random mud lessened the effectiveness of ordnance in all theaters of the war. Officers and infantrymen wrote of shells being entirely swallowed by mud on the Mississippi bird-foot, Virginia peninsula, and Georgia barrier islands. In a harsh fight for a coastal or riverine fortification, this diminishment of artillery firepower could be a critical factor in the success or failure of an operation.

While the clay in mud is the smallest particle size that made warfare difficult and soldiers' lives miserable, a clastic particle orders of magnitude larger also had an impact on the fighting: cobbles and boulders. The most famous of these incidents occurred along the unfinished railroad grade at Second Manassas when the rebels ran low on ammunition.[3] What better way to illustrate the most basic relationship between geology and combat: the Stonewall Brigade—members of Stonewall Jackson's Second Corps—hurling stones to defend Stony Ridge.

During the final stages of Lee's Maryland campaign, cobbles were again used, this time by desperate soldiers to fire their new, but defective, Enfield rifles. Members of the 118th Pennsylvania approached a little too closely to the retreating Army of Northern Virginia's rear guard and were attacked by five thousand men of A. P. Hill's division. Here, trapped against the cliffs above the Potomac, the Pennsylvanians found the mainsprings of their British rifles too weak to explode a percussion cap. While most men discarded their malfunctioning weapons, a significant fraction turned to limestone and sandstone cobbles to explode the caps, discharging their guns.[4] This no doubt led to some highly accurate, long-range rifle fire.

Confederate soldiers used all types of large sedimentary clasts during the fighting around Kennesaw Mountain during Sherman's campaign to capture Atlanta. Cobbles rained down on unfortunate Federal soldiers during the fierce combat at the Dead Angle at Cheatham Hill, and migmatite boulders rolled down the steep face of Pigeon Hill and Little Kennesaw Mountain on more than one occasion.[5] This fighting certainly represents the largest grain-size weapon used during the war.[6]

Sand and Postbellum Coastal Fortifications

The permanent works of Forts Pulaski, Sumter, Macon, Jefferson, Jackson, Morgan, and Caswell all belonged to the Third System of Fortifications.[7] These attractive masonry structures were constructed between 1816 and 1867.[8] Gillmore's capture of Fort Pulaski and reduction of Fort Sumter clearly documented in rubble the vulnerability of brick to large-caliber, land-based, rifled artillery. The prevailing thinking entering the Civil War had been that the primary threat to these forts would come from the sea, and thus their firepower was concentrated toward the water. Simply put, the forts were designed to overwhelm any approaching enemy vessel with cannon fire from multiple levels of casemates before any substantial damage could be suffered. This strategy proved successful in the case of the failed Federal monitor attacks against Fort Sumter in 1863.

To protect the fortification from a land attack, many forts had outworks to shield the brick ramparts from direct cannon fire. Outworks were commonly constructed of earth. At Fort Pulaski, for example, a large earthen ravelin protected the land face of the fort and moat (see figure 20.1). The sea face was left unprotected by sediment; with the closest hard ground located more than a half mile away on Tybee Island, the fort was mistakenly thought to be invulnerable to long-range enemy fire. The key to destroying the capacity of both Pulaski and Sumter was the employment of land-based rifled artillery, and in the case of the Georgian fort, to attack the sea face from land.[9]

FIG. 20.1. To enter Fort Pulaski's sally port you must cross two moats. The first allows entrance to the ravelin (demilune, right), the second crosses the larger primary moat (center) that surrounds the fort. The ravelin complicates access to the sally port for the enemy and protects the brick scarp from direct land-based artillery fire.

Gillmore was clear in his after-battle reports that, in the future, all masonry fortifications should be modified to incorporate the benefits of sand.[10] His engineering reports praise clean, mature quartz sand for use in fortifications: "Against the destructive effects of projectiles, pure quartz sand, judiciously disposed, comports itself unlike any other substance, and for certain parts of fortifications its particular properties suggest its exclusive use, in preference to ordinary earth."[11]

Some Civil War–era fortifications, like Fort Moultrie on Sullivan's Island at Charleston or Fort Caswell at the mouth of the Cape Fear River, had already been modified with sand piled against the exterior of the ramparts for protection.

Larger, faster, steam-powered ships with substantially heavier armament became an even more sizable menace to the Third System forts in the years after the Civil War. To deal with this increasing threat, extremely heavy defensive artillery was needed, and the emplacement of 15-inch Rodman guns required the redesign of many forts and the construction of new outworks. The shift in the coastal defense philosophy by the American military was succinctly described by fortifications expert John Weaver: "The weapon became the item of primary importance in seacoast defense, rather than the fortification that housed the weapon."[12]

In Europe during this same time period there was an explosion, so to speak, in the exploration of the use of iron in coastal defensive works. Unlike some wealthier European powers, the cash-strapped United States could not afford to add extensive iron plating to their fortifications, not to mention the upkeep of the saltwater-intolerant material. As a result, the focus of fortification design and construction for the defense of the American coastlines remained tied to sediment.[13]

As the United States approached the turn of the twentieth century, the effectiveness of sand parapets in coastal fortifications was not forgotten. In 1896 the U.S. Board of Engineers began comparing sand to the newest material used in coastal fortifications: concrete. They calculated that one foot of concrete had the penetrative resistance of 2–3 feet of sand. As improvements in the chemistry of concrete occurred over the next decade, this ratio was increased until it was closer to 1:3. Other military theoreticians of the time argued that these ratios underestimated the power of sediment, especially at deflecting low-trajectory incoming fire and smothering explosive shells.

In his 1920 *Notes on Seacoast Fortification Construction*, Col. Eden Wilson detailed the benefits of sand over concrete. Sand, he argued, was both more effective and less expensive: "The same amount of protection could be obtained at less cost with a comparatively thin concrete wall, having sand in front of it, than if concrete was used throughout." Wilson further described how sand "serves to blanket the explosion and reduce the scattering of fragments" when compared with concrete after parapet penetration and that, unlike solid materials, there are no resulting "exceedingly dangerous" secondary projectiles.[14]

Nevertheless, many planners doubted how a sand parapet could be expected to withstand the impact and explosion of modern (World War I–era) 12-inch rifled artillery firing shells at more than 2,000 feet per second. Wilson recommended building the reinforced, thick sand parapet far in advance of the seacoast artillery and piling sand well below the angle of repose. Impact from enemy shells would essentially only move the parapet around, far away from the protected defensive artillery position. The only downside of this construction, however, was the requirement for indirect fire by the seacoast artillery: if the enemy can't directly see and fire at your position because you are hidden behind a massive pile of sand, you can't see the enemy and directly return fire. Artillery shells also became more streamlined and aerodynamic around this time. This change allowed them to retain velocity and energy at greater distances and penetrate sediments more effectively and efficiently.

By the end of the nineteenth century, new and extremely powerful breech-loading steel artillery completed the obsolescence of the brick Third System forts. New coastal fortifications were constructed of concrete, with little concern for landward-facing defenses. Sand would nevertheless continue to play a critical role in defensive works. Historian Emanuel Lewis summarizes the equation that kept sand and soil from becoming as archaic as brick in coastal fortifications as technology improved: "The inability of masonry to withstand modern weapons, the postwar shortage of funds for military purposes, and the need for emplacements large enough to receive the new armament combined in the closing years of the decade to bring about a return to an inexpensive mode of permanent fortification in which earth again became the principle substance of protection."[15] This new series of fortifications and batteries would be called the Endicott System, named for the secretary of war. Widely dispersed and hidden long-range guns, huge concrete emplacements, and underground magazines provided enhanced shoreline defense to the United States as they entered first the war with Spain and later World War I.

Steel and concrete were the new preferred construction materials for defensive works, but sediment remained a key component of the protection, if for no other reason than because sand is so inexpensive (see figure 20.2). J. E. and H. W. Kaufmann describe the concrete/sand relationship:

> To cut costs in these batteries, sand replaced concrete as fill in certain areas such as between floors. A concrete blast apron in front of the parapet where the guns fired prevented the earth cover from being blown away during firing. According to the Board of Engineers report of 1903, either 45 feet of concrete or 90 feet of sand were required as protection against the heaviest naval guns based on tests of Rosendale cement.[16]

Despite all of these improvements to the construction materials used in coastal fortifications, one important attribute of sand remained unchanged: fortifications constructed from sand can be rapidly repaired

FIG. 20.2. Battery Langdon guarded Pensacola Harbor with 12-inch guns. The entire battery, except for the embrasure, was buried under several feet of Gulf-Coast sand. Battery construction began in 1917 and completed in 1923.

and reconstructed if damaged by enemy shellfire. No other material used in construction—not steel, concrete, nor brick—can be as quickly repaired, and this attribute of defensive building material doesn't become apparent until the structure is sustaining damage.

Lessons Learned for the Great War

Evolution of fieldworks was significantly slower than that of small arms and artillery technology during the second half of the nineteenth century. Advances in metallurgy led to the creation of more powerful and longer-ranging artillery, and the introduction of bolt- and lever-action repeating rifles and rifle cartridges changed both the nature of combat and the types of fieldworks that were effective and could survive on the battlefield.

By the start of the First World War, field artillery was mobile enough and powerful enough to prohibit use of taller earthen parapets such as those created during the Civil War. Parapets made of sand or soil were seemingly penetrable regardless of how thick or well constructed, largely because of advancements in chemistry allowing smokeless gunpowder to propel new forms of more explosive shells at higher velocities over longer ranges. Most parapets were too high and vulnerable to this artillery even if they featured interior ditches. Muzzle velocities first doubled and then nearly tripled during these decades, and fuses

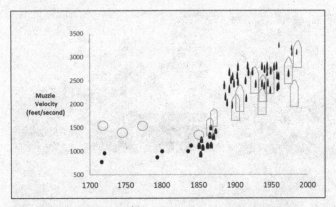

FIG. 20.3. Increase in muzzle velocity of primary infantry and sharpshooter rifles (black, solid) and field and heavy artillery (gray, open) 1750–2000. Note that average muzzle velocity for infantry weapons essentially doubled during the two decades following the American Civil War, and artillery muzzle velocities increases by 500–1000 feet/second.

became more reliable and predictable. The result of these improvements in firepower was a need for smaller, thicker, lower parapets and a deeper interior ditch: the trench of World War I.

The design, construction, digging, and revetting of trenches from the First World War was in many ways similar to those from the Civil War fifty years earlier. The primary difference between the fieldworks involved their visibility to the enemy: World War I trenches provided almost no target while Civil War parapets and gun embrasures would have been highly vulnerable to World War I artillery or rifles.[17] A good sniper in 1917 could easily fire effectively at a gun crew in an embrasure from more than 400 yards away. This range was possible because of the dramatic increase in muzzle velocity as armies moved from the musket to the small-bore bolt-action rifles firing smokeless powder (see figure 20.3). Muzzle velocities exceeding 2,500 feet per second equated to much flatter trajectories and equally impressive increases in effective range.[18]

Changes in small arms were responsible for the origin, evolution, and demise of Civil War fieldworks. During the era of the smooth-bore musket, defenders could feel relatively secure beyond around 200 yards, and the need for field improvements was deemed somewhat less than critical. If anything, entrenching was thought to lower the offensive aggressiveness of infantry. The shift to the new, deadlier, longer-ranging rifle musket in the 1850s extended the range of vulnerability. Well-trained infantry could hit an individual target at 200 yards and beyond, and formations of men could be consistently struck at twice that range. When small-bore repeating rifles began to be used in quantity on the battlefield, the firepower of infantry increased exponentially. German Mausers and British Lee-Enfield rifles fired a full metal jacket bullet half the diameter of a Springfield rifle musket at more than twice the velocity. Equally important, Great War rifles could be fired and reloaded in a fraction of the time it took to reload the clumsy musket, and a soldier could do so while lying on the ground. Advances in artillery, in terms of increased firepower and range and the rapidity of breech reloading, were equally dramatic. All of this meant the outmodedness of nearly all field improvements that were aboveground and could be penetrated or blasted, including earthen parapets and breastworks. It also meant the end of massed infantry attacks over open terrain such as on the third day at Gettysburg.[19] How many well-armed and supplied World War I infantry soldiers would it have required to repel Pickett's Charge if they were positioned on Cemetery Ridge on July 3, 1863? Perhaps a few regiments—a couple thousand men? Accurate defensive fire would commence as soon as the Confederates emerged from the tree line on Seminary Ridge with no fear of reducing visibility from black-powder smoke. As Lee's infantry marched across the gently sloping terrain underlain by sandstones and shales, they would have been annihilated—and this imaginary scenario does not even incorporate the presence of machine guns or World War I artillery.

After the first year of World War I, the combat began to resemble the siege lines around Petersburg and Richmond in 1864, but on a much larger scale. With flanking options growing impossible and offensive tactics proving more and more costly, the temporary trenches grew more complicated and permanent. Mines began to appear under the battlefield as well.

The geology of many World War I battlegrounds was more similar to the carbonate battlefields of Tennessee (Chickamauga, Stones River, Nashville, Franklin) or Maryland (Antietam, Monocacy) than Richmond or Petersburg. Nevertheless, forest cover was more prevalent in North America, providing concealment for infantry movement. At Stones River, for example, significant portions of the battlefield have outcropping limestone or thin topsoil above the bedrock, prohibiting tillage. These areas were left as forests and cedar glades, and areas with thick tree cover necessitated entrenching to defend against sneak attacks by the enemy. Entrenching was impossible with shallow bedrock, however, so the resulting outcome was a natural geological prohibition on digging in an area where it was most desirable.

Ypres was not located above carbonate rock. Instead, this battleground was more similar to the Coastal Plain battlefields of Petersburg or Fredericksburg. As such, the trenches cut into the northern European soil had many of the same problems as those dug in Virginia: much clay brings drainage problems, and much sand leads to slumping that requires extensive revetting.

On the farm fields of Flanders or the Somme, the soil development was mature and deep enough and the consistently eroded limestone and chalk landscape was flat enough to encourage massive entrenching efforts. It is then surprising that, much like the Civil War, entrenching was not immediately ubiquitous during the first year of the Great War.[20] At least part of the reason for this lesson not learned has more to do with geography than geology. European military academies used case studies from nineteenth-century European battles, not fighting across the Atlantic.[21] The combination of changing weapons technology and the fifty years between the American Civil War and the Great War appear to have diminished the Europeans' consideration of how field fortifications changed between 1861 and 1865 and what these lessons might portend for battles on a much larger scale.

Geologists played a different role in World War I than they would in later conflicts, in part because of the static nature of the warfare.[22] During the First World War, geologists were tasked with four primary responsibilities: finding suitable water supplies, evaluating the landscape and soils for the suitability of entrenching, locating supplies of strategic minerals, and identifying quarrying sites for aggregate for road construction.[23] Without extensive experience in military geology and engineering, geoscience-based tactical mistakes were common. Trenches were dug during droughts with no thought of the water table, only to become elaborately constructed canals a week later. Months were spent tunneling toward the enemy line only to have the shaft strike igneous or metamorphic rock, leading to immediate abandonment of the project. Road aggregate was transported with great but unnecessary effort across the English Channel from southern England when ample supplies of the material lay undiscovered in nearby Brittany.[24]

Sedimentary Geology in Later Conflicts

Sand and sedimentary geology were studied from a military and scientific perspective for the first time during the Second World War. This was a much more mobile war, bringing new challenges to geologists. Concrete began to replace sand in many defensive applications, at least to some extent, and long lines of elaborate earthworks became less important than the seemingly infinite number of simple foxholes. More efficient water-purifying techniques meant that surface water could often supply the needs of an army in the field. Geologists could spend the time they would have been drilling groundwater wells on other engineering projects, like finding a suitable topographic location for a new airfield.[25]

Geologists gained innovative tools during the final years of the war. High-resolution aerial photographs of bomb craters, for example, were used to reveal the previously hidden subsurface geology of strategically important, yet unmapped, regions of Europe.

Sedimentary geology came to the forefront in importance during this war because of the fighting along, and for, the shorelines. World War I had witnessed the disastrous amphibious landing of the British empire's forces at Gallipoli, but the Second World War had war-changing shore-face landings on three different continents. Sand and sediments played an important role in the fighting for the beaches from the white quartz sands of Anzio to the black pebble and cobble basalts of Iwo Jima. The most important of these amphibious landings, along the Normandy coastline, involved sedimentary geologists from the inception. Supreme Headquarters of Operation Overlord was concerned that the beaches along the Calvados coast would not be able to support trucks and tanks. Aerial reconnaissance revealed great quantities of soft peat along the shore face—an indicator that an underlying and largely hidden mud might exist in quantity to trap vehicles on the beaches. Thus, the bearing capacity of the beaches, and the likelihood that tanks, bulldozers, heavy trucks, and DUKWs could escape this deadly portion of the shorefront became a matter of critical concern.[26]

To quantify the significance of this sedimentary threat to the proposed amphibious landings, a team of British commandos was commissioned to take sediment samples and short auger cores from the shoreline that would later be code-named Sword Beach. On New Year's Eve 1943, a landing party of Maj. Logan Scott-Bowden of the Royal Engineers and Sgt. Bruce Ogden-Smith of the Royal Navy fought tides and waves to capture sand and mud. Their sampling equipment included 10-inch tubes for bulk sediment samples and an 18-inch auger for extracting a series of short cores. Earlier research, based in part on car racing at Daytona Beach, Florida, had established a general rule that at least 14 inches of sand was needed to support vehicular traffic leaving the water.[27] As a result, the 18-inch sediment cores established confidence in the Allied planners that the beaches at Normandy had, at least in most sectors, the sandy beaches that could support tank traffic. Scott-Bowden and Ogden-Smith landed again on the beaches on June 6, making them the only men to land along the Normandy coast more than once during the war.[28]

The amphibious landings in North Africa, Italy, France, and the South Pacific gave military engineers and geologists a long list of considerations for future joint army-navy operations. Landing sites that previously would have been considered impracticable, if not impossible, for a successful invasion were now considered viable. In the Korean Conflict, for example, the harbor at Inchon had treacherous currents that approached 10 mph and a ridiculously large tidal range of more than 30 feet.[29] At D-Day in France the landings took place at low tide to expose beach obstacles. At Inchon, low tide exposed vast mudflats, making a landing impossible. As a result, the assault was planned for a spring high tide, which would bring the landing craft ashore more than 10 feet above the flats, to a much narrower and steeper beach. This plan, however, dictated the timing of the second wave of supporting landing craft; they could not reach the shore until the next high tide, more than twelve hours later. The stakes had been raised for a successful initial landing by the sedimentary geology, tidal regime, and small size of the harbor.

Sedimentary geology played an entirely different role in the next major conflict in eastern Asia. Seasonal muds and sandbags seemed to define the nature of the conflict in Vietnam. The Civil War origins of the simple firebase are easy to observe on the tropical hills and mountaintops: rectangular and star-shaped parapets, lunettes, and redans capped with sandbags. The clay-rich nature of the soil led to tunneling on a grand scale, and the ever-present mud limited the effectiveness and deployment of American armor: only 46 percent of the country's soft landscapes could support tanks.[30]

While the seasonal and random mud of Vietnam diminished the effectiveness of tanks, the intergrain friction provided by the sands of the Middle East turned the M1A1 Abrams tank into an especially potent offensive weapon during the Gulf War. Of all the military conflicts from the nineteenth and twentieth centuries, those of Operations Desert Shield and Desert Storm in Iraq, Kuwait, and Saudi Arabia are the most easily associated with sedimentary geology. Sand was everywhere—on and under the battlefield, in the air, on and in the vehicles—even in the rain. Around one-third of the Arabian Peninsula is covered in mobile sand.[31]

The terrain is generally flat, with the primary geologic features in some areas being 15–30-foot-high sand dunes and rounded hills. Tank movement and transport was primarily hindered by migrating sand dunes and drifts (*dikakah*) and *sabkhahs*—unpredictable saline flats underlain by clay, sand, or silt.[32] Wadis, or dry river channels underlain by cobbles and pebbles, were especially good terrain for an armored attack.

The sand created terrific problems, however, for all kinds of mechanical equipment. Sand could foul an M16 as easily as a Springfield rifle musket. Artillery was also plagued by the grit. Soldiers on Morris Island in 1863 had complained bitterly of airborne sand and the havoc it caused; imagine the situation during a full-blown Middle Eastern sandstorm.[33]

Of all the equipment most vulnerable to the abundant, mobile, fine-grained sand, one stands (flies) above all others: the helicopter. Sand can erode the leading edge of rotor blades and rotor heads, and exposed flight surfaces can be destroyed quickly through abrasion.[34] Sand can also be ingested into engines causing failure, and grit can cause machine guns and cannon to seize. Airborne sand can also disorient the flight crew, and a helicopter taking off or landing is constantly providing its own aeolian force.

During the Battle of Petersburg sand was thrown as a last resort by soldiers under attack in the trenches in a desperate attempt to blind the enemy. In the Gulf War sand was used as an offensive weapon on a much larger, and more controversial, scale. The Iraqi army had prepared an enormous and compound trench system in the sand known as the Saddam Line. The U.S. First Infantry Division (Armored), faced with this defensive network, chose to use the trenches against the defenders, sending forward tanks with mine-clearing plows and armored earthmovers to collapse the earthworks and bury the Iraqis alive.

Modern Exploitation of Sediments on the Battlefield

The widespread use of simple sandbags is the most enduring application of sediment to solve a military problem that emerged from the Civil War. Sandbags had been employed in earlier conflicts—Loyalists used them when defending Fort Ninety Six in the American Revolutionary War, for example—but parapets, embrasures, and trenches lined with thousands or even tens of thousands of bags were first seen during the Civil War.

The tactical use of sandbags has changed little between the Civil War and the recent strife in Afghanistan, but the material used for their construction and the methods used to fill the bags have changed significantly. For example, during the Charleston Campaign in 1863, the Federal army used nearly fifty thousand sandbags.[35] Before being filled these bags were made of cloth and measured 6 by 10 inches and were 24 inches long.[36] Once filled by hand with local sediment, they weighed approximately 85 pounds, although the bags were rarely stuffed to capacity; completely filling the bags stretched the cloth, inclining the sandbags to tear and leak.

Today, the U.S. military uses sandbags that are often filled by machines. Modern sandbags are larger than those from the Civil War, measuring 14 by 26 inches, and are made of treated and plain burlap, polypropylene, and acrylic. For nonmilitary applications (e.g., flood control or levee repair) the U.S. Army Corps of Engineers suggests filling bags with sand, and only sand, whenever possible.[37] Silt is difficult to shape, according to the Corps, and the fine-grained material may escape through the weave in the bag. Clay is also difficult to shape and bag.[38] Coarser materials such as gravel lack the impermeability to make the bags effective for water control.

Sandbags are not the only revetment material from the Civil War to have a direct descendant on the modern battlefield. Gabions, essentially wicker baskets that were open on both ends and filled with sand or soil, were used extensively during the Civil War. Gabions could be used as revetments, supporting larger parapet walls, or they could stand alone as smaller earthworks or traverses. Gabions share two advantages with sandbags. First, they are relatively easy to transport because, when not filled with sediment, they are light in weight. They are also relatively low-maintenance once emplaced and filled. Just as with

FIG. 20.4. HESCO barrier surrounding an M2 Bradley Fighting Vehicle outside the U.S. Army Heritage and Education Center in Carlisle, Pennsylvania. These gabions are filled with coarse sand, although any material ranging from silt to gravel can be used to provide protection. When empty, these modern gabions can be easily moved by a single soldier; when filled with sand they weight around a ton.

sandbags, the filling material must be of appropriate coarseness: sediment larger than silt is required in sandbags and sediment larger than fine sand is used in most gabions, or the fill will seep out.

Gabions exist in defensive positions on modern battlefields as well, albeit with the replacement of wicker and small branches with more durable (and leakproof) material. One modern gabion that has seen extensive use in Iraq and Afghanistan is the HESCO MIL, a defensive device developed in the late 1980s (see figure 20.4). The HESCO system consists of a collapsible steel wire mesh lined with a polypropylene geotextile. These structures offer more protection than the woven basket-of-branches gabions from the Civil War, while also being simpler to assemble and fill. A single gabion from the 1800s might take a day to assemble and an hour to fill; with proper heavy equipment, an entire traffic control point or forward operating base can be fortified with modern "gabions" in less than a day.

Military geology was a new concept for engineers and theorists during the American Civil War, and the understanding of the exploitation of terrain for military purposes was evolving on several scales. These ranged from the movement of entire armies hidden behind mountain ranges to the positioning of a single soldier in a freshly dug foxhole. A legacy for future wars was created by advances in earthwork engineering between 1861 and 1865, and these improvements were well tested in battle. The walls of earth, or parapets, thrown up in front of ditches marked the armies' positions on the battlegrounds after the fighting moved elsewhere, and many survive today, having avoided erosion and urban development. Nevertheless, the tactical use of 5-foot-high parapets did not survive far into the First World War. By the turn of the twentieth century, the increased firepower of artillery necessitated that only the flattest, thickest parapets in front of trenches would withstand bombardment. The increased explosiveness of artillery shells during World War I diminished the effectiveness of earthen fortifications to the point that vertical structures could only be dug, not piled. "Safety," if it was to be found, was *in* the earth, not behind it. *This leaves the Civil War as the most suitable time period in military history to analyze the relationship between sedimentary geology and infantry tactics.*[39] By 1914, artillery, aircraft, and fast-firing bolt-action rifles and machine guns changed the role of the military engineer and the impact of local geology for all future wars.

Sedimentary geology remained a subject of interest during both world wars, and it reached its zenith when British engineers landed covertly on the beaches of Normandy to take samples of beach sand in the weeks prior to Operation Overlord.[40] From the bottomless mud of Vietnam to the sandstorms of the Gulf War, sediment continued to play a major role in shaping the fighting and outcome of battles.

Military Conflicts as a Geological Tool

This book has focused on the influence of sediments and sedimentary geology on the combat, tactics, and strategy of the American Civil War. A new field of study has emerged during the last decade that flips this line of inquiry: instead of discussing how geology influenced war, the research brings attention to how war affects geology.

Instead of bioturbation, the mixing of sediments by burrowing creatures, we now have bombturbation, the science of how the evolution of the landscape is changed by blasting it to hell. Multiple comparative

FIG. 20.5. A bomb crater created in the spring or early summer 1944 at Pointe du Hoc. This crater lies less than 20 feet from the cliffs overlooking the Calvados Coast. The U.S. 8th Air Force, British Bomber Command, and the USS *Texas* battleship all contributed to the bombturbation of this site.

studies have documented changes to the battleground landscape in the post-conflict decades: How has cratered terrain responded to changes in surface runoff on soil development? Hupy and Koehler, for example, compared the geographically dissimilar obliterated landscapes of Verdun, France, and Khe Sanh, Vietnam. As good—and perhaps overly descriptive—scientists, they refer to an exploding shell or bomb as a "zoogeomorphic disturbance."[41] Later, tubular subterranean annelids (earthworms) dig through the craters, creating an interesting battle between bombturbation and bioturbation (see figure 20.5).

This zoogeomorphic disturbance is valuable to geologists because the timing of the event is known precisely: it acts as a natural, highly visible, unmistakable tracer left in the sedimentary record.[42] Another even more valuable tracer in the sedimentary record is also created by the explosion of a bomb, although this time the weapon must be of the nuclear variety. The isotope cesium-137 is the by-product of nuclear weapons testing (and use). The radioactive particles of this isotope are spread throughout the earth's atmosphere after an explosion, where they reside for approximately one year before settling to the surface of the planet.

When cesium-137 spreads across a sedimentary depositional environment, it will be incorporated into the strata. Marshes, lagoons, and lakes are all good areas for this radiotracer to accumulate in sediments. The concentration of the radiotracer is proportional to the amount of cesium in the atmosphere, which is proportional to the number of weapons being detonated.[43] Sediments deposited before 1946, the first year of deposition after the development of the bomb, are bereft of cesium. The year 1962 marked the peak of nuclear weapons testing in the atmosphere, and so the next year, 1963, was the year of maximum deposition of the radiotracer. Sediments deposited in 1963 show a higher concentration of cesium-137—a spike in the abundance in the sediment. This cesium spike is a valuable geochronologic tool to earth scientists because it precisely defines the time of the layer of sediment's deposition.[44]

One modern application of cesium-137 is especially valuable in our current era of dramatic climate and ocean change. Salt marshes and salt-marsh vegetation are found between the elevation of high tide and low tide. Saltwater-tolerant plants (halophytes) prefer to be in the marine water during only one portion of the day; thus the surface of the marsh is a good proxy for the surface of the ocean (i.e., sea level), and buried marsh sediments and halophytes represent former elevations of sea level. The cesium-137 spike can be used to identify the elevation of the oceans a half century ago, and as a result, the rate of sea-level rise can be reconstructed, even for remote areas that lack tide gauges or measures to record previous sea-level elevations. If, for example, the cesium spike is found in strata buried two feet below the salt marsh surface, the radiotracer indicates that this was the level of the marsh (i.e., sea level) in 1963. This region had a two-foot rise in sea level since 1963 (0.4 inches/year).[45]

Radiotracers were even used as a time-marking or geochronological tool for the sediments from inside the *H. L. Hunley*. Because the muds and sands from all layers of strata within the hull were cesium-dead, the timing of the infilling must have occurred prior to 1945.[46]

A century in the future, American Civil War battlefields will inevitably and undoubtedly have changed. Landscape features, those both natural and created by the combatants and slaves of the 1860s, will continue to weather and transform. Some battlefields and earthworks will surely be lost to wave erosion. Permanent fortifications along the Atlantic and Gulf coasts will be especially vulnerable to the dual threat of rising sea level and storm erosion. Future threats to battlegrounds, whether from anthropogenic or natural sources, are difficult to foresee.

The nature of future warfare is even more difficult to predict than the fate of the Civil War battlefields. Smarter, more efficient technology and more accurate and powerful weapons will likely be used in these hopefully distant conflicts. Nevertheless, one important defensive tool from the Civil War will still be used and remain largely unchanged: the widespread use of sand in bags. The material of the bag will be different, but the sediment will not: medium and coarse quartz sand. It is difficult to imagine another weapon, tool, or military apparatus that has remained so consistent over centuries of use, and the influence of the humble sandbag is entirely related to the nature of the fill material, not the textile retaining the sediment.

CHAPTER 1. Sediments and Conflict

1. Most of the existing literature dealing with the relationship between sedimentary geology and the American Civil War has focused on soil. For example, Harold Winters dedicated many pages of his comprehensive work *Battling the Elements: Weather and Terrain in the Conduct of War* (1998; John Hopkins University Press) to the formation and impact of mud on fighting throughout the history of warfare. More recently, Lisa Brady's *War upon the Land: Military Strategy and the Transformation of Southern Landscapes during the American Civil War* (2012; University of Georgia Press), Brian Allen Drake's *The Blue, the Gray, and the Green: Toward an Environmental History of the Civil War* (2015; University of Georgia Press), and Judkin Browning and Timothy Silver's *An Environmental History of the Civil War* (2020; University of North Carolina Press) have taken a broader view, applying an environmental history approach to the war and its aftermath. These texts deal indirectly with sedimentary geology, discussing depositional environments and artificial modifications to the landscape that were the direct result of the fighting on and around the battlegrounds.

2. Stephen W. Sears discusses the rapidity of transition between dust and mud during Maj. Gen. George B. McClellan's 1862 Peninsula Campaign against Richmond: "After two or three days of hot sun the roads would be dry and turn to thick powdery dust; after an hour's rain they turned back to bottom-less mud." (1992, *To the Gates of Richmond: The Peninsula Campaign*, Boston: Ticknor & Fields, 468p., quotation on p. 108).

3. Sediment grain size even influenced the manner of burial for soldiers after battle. At Gettysburg, those men killed around the igneous rock Round Tops were hastily buried under piles of diabase cobbles and boulders, while soldiers interred less than a mile away were buried underground in pits cut into the softer sedimentary strata. Browning J, Silver S, 2020, *An Environmental History of the Civil War* (Chapel Hill: University of North Carolina Press), 272p.

4. These figures may seem hyperbolic, but they are not. One cubic foot of sand would contain around a billion grains (depending on the grain size). A 100-pound bag of fine sand would also contain around a billion grains of sand.

Considering the size of sand fortifications like Fort Fisher along the North Carolina shoreline, it is easy to see how more than a quadrillion grains of sand would be required for construction.

5. These investigations of geology and fighting were first discussed in the literature about the Battle of Gettysburg during the centennial of the war. One of the first was published by the Pennsylvania Geological Survey: Brown A, 1962, *Geology and the Gettysburg Campaign: Pennsylvania*, Topographic and Geologic Survey Educational Series no. 5, pp. 1–15. More recently, the subject has been revisited by Inners JD, Cuffey RJ, Smith RC II, Neubaum JC, Keen RC, Fleeger GM, Butts L, Delano HL, Neubaum VA, Howe RH, 2006, Rifts, Diabase, and the Topographic "Fishhook": Terrain and Military Geology of the Battle of Gettysburg—July 1–3, 1863, rev. ed., Pennsylvania Topographic and Geologic Survey Open-File Report 06–02, 111p. The campaigns for Second Manassas and Chickamauga are discussed by Zen E, Walker A, 2000, *Rocks and War: Geology and the Civil War Campaign of Second Manassas* (Shippensburg, Pa.: White Mane), 109p.; and Henderson SW, 2004, "The Geology of the Chickamauga Campaign, American Civil War," in Caldwell DR, Ehlen J, Harmon RS, eds., *Studies in Military Geography and Geology* (Dordrecht, Netherlands: Kluwer), pp. 173–84, respectively.

6. Only the combination of hard igneous diabase (a rock similar to basalt) and softer sedimentary sandstones and siltstones around Manassas, Virginia, are comparable. The rocks in Virginia, however, do not crop out as dramatically as they do at Gettysburg. As a result, geology at the Battle of Second Manassas had a subtler influence on the fighting.

The Gettysburg battlefield might also be considered one of the birthplaces of combat photography as Thomas H. O'Sullivan and Alexander Gardner captured many of the most famous and gruesome photographs to emerge from the war in the days after the fighting ceased.

7. Stratigraphy is the study of layered rock. For example, tilted rock layers often produce rolling terrain, while flat, horizontal strata results in a lack of relief.

8. Ehlen J, Whisonant RC, 2008, "Military Geology of Antietam Battlefield, Maryland, U.S.—Geology, Terrain, and Casualties," *Geology Today*, v. 24, pp. 20–27.

9. Most of the casualties were from concussion.

10. This attack was depicted in the climax of the film *Glory* (directed by Edward Zwick, 1989, TriStar Pictures).

11. In many cases, after brick fortifications were blasted into rubble, the brick fragments, now an anthropogenic sediment, allowed the reduced fortification to retain a surprising degree of its former defensive capabilities.

12. For a thorough discussion of the attitudes of military leadership regarding the need or desire to have their soldiers entrench, see Hess EJ, 2013, *Field Armies and Fortifications in the Civil War: The Eastern Campaign, 1861–1864* (Chapel Hill: University of North Carolina Press), 448p.

13. Sappers were combat engineers.

14. McPherson was describing Fort Fisher, comparing the effects of artillery upon this sand fort to the brick Fort Sumter in *Battle Cry of Freedom* (2003; Oxford University Press).

CHAPTER 2. The Coevolution of Military and Geological Sciences

1. Mahan authored a number of important books, yet his own career at West Point and the influence he had on the future generalship on both sides of the Civil War have never been explored in detail in a full-length text.

2. Lyell's *Principles of Geology* greatly influenced Charles Darwin during his travels on the HMS *Beagle*, which finished its voyage the same year (1836) that the military world was introduced to Mahan's *Treatise*.

3. The study of the shape and evolution of the earth's surface is a branch of the geosciences called *geomorphology*.

4. Lyell was looking for fossils in Georgia that might allow him to correlate the North American rocks with those from Europe. For more on the environmental aspects of his journey, including his writings about the degradation of southern soil, see Timothy Johnson's chapter "Reconstructing the Soil: Emancipation and the Roots of Chemical-Dependent Agriculture in America," in Drake BA, ed., 2014, *The Blue, the Gray, and the Green: Toward an Environmental History of the Civil War* (Athens: University of Georgia Press), pp. 191–208.

5. Mahan also traveled across the Atlantic, spending four years in France where he studied the French military engineers' use of fortifications and entrenchments.

6. One recent example of the intersection of the natural (physical) sciences and military history was contributed by two authors with disparate backgrounds: Judkin Browning, a professor of military history, collaborated with Timothy Silver, a professor of environmental history, to produce *An Environmental History of the Civil War* (2020, Chapel Hill: University of North Carolina Press).

7. For a detailed treatment of the influence of climatology on the landscapes of the Civil War, refer to Browning and Silver's *Environmental History of the Civil War*. While several authors have addressed the role of weather on particular battles or campaigns, Browning and Silver explore how larger-scale climatic shifts altered everything from disease vectors to agricultural productivity.

8. A clastic sediment is made of detrital materials (sand, silt, pebbles) and becomes lithified when it is cemented and compacted to become a rock. Thus, a sand is lithified to become sandstone, silt is transformed into siltstone, and pebbles are cemented together to become either conglomerate (round pebbles) or breccia (angular pebbles).

9. Salt, in the form of halite, is also classified as a chemical sedimentary rock. Mark Fiege (2013, *The Republic of Nature: An Environmental History of the United*

States; University of Washington Press) discusses how the "salt famine" proved progressively more of a problem for the Confederacy as the war continued.

10. These rocks have been radiometrically dated to around 200 million years old (Olsen PE, Kent DV, 1999, "Long-Period Milankovitch Cycles from the Late Triassic and Early Jurassic of Eastern North America and Their Implications for Calibration of the Early Mesozoic Time-Scale and the Long-Range Behavior of the Planets," *Philosophical Transactions of the Royal Society of London: Mathematical, Physical and Engineering Sciences* series, v. 357, pp. 1761–86; Smith RC 2nd, Keen RC, 2004, "Regional Rifts and the Battle of Gettysburg," *Pennsylvania Geology*, v. 34, pp. 2–12.

11. One of these fractures was responsible for the 2011 earthquake near Mineral, Virginia, that surprised so many people across the mid-Atlantic states.

12. As well as the rocks underlying the famous unfinished railroad cut at the Battle of Second Manassas.

13. This is an example of the stratigraphic principle known as *cross-cutting*: the rock that is being cut is older than the rock that is cutting across it.

14. The Fall Zone is also often referred to as the Fall Line, although it is usually not entirely linear in nature or easily traceable across the landscape.

15. The influence of the Fall Zone on Civil War logistics is explored in more detail in chapter 8.

16. One aspect of quartz sand that is often overlooked, from a military perspective, is the discomfort the bright color of the sediment can cause for soldiers stationed along the shoreline. For example, Brig. Gen. J. W. Phelps suggested that his men, who were occupying Ship Island, Mississippi, might require some sort of eye protection from the glare of the white sand during the summer months. In a report from January 1862, he suggested that he might need to relocate his camp because of the hot and blinding white sand, "if the rebellion should last through the summer." *The War of the Rebellion: A Compilation of the Official Records of the Union and Confederate Armies*, ser. 1, v. 6, quotation on p. 688.

17. Weathering is the breakdown of rocks in place. The term "erosion" represents the transport of these weathered particles to a new location.

18. The best example of a "mature," clean sand is abundant along the coastlines: most beaches and nearly all dunes contain sediment that meets these criteria.

19. These sedimentary rocks are also a tremendous reservoir for carbon, capable of storing the element for millions of years. A test is an internal shell, found in sea urchins, sand dollars, and foraminifers.

20. "Major battle" defined as one of the top twenty with respect to the number of troops engaged.

21. On the Mohs' scale of hardness, limestone is a rather soft "3" and dolostone is a "3.5–4." Chalk would be softer, and quartz-sandstone, much harder.

22. For a review of the impact of carbonates during this fighting and an interesting application of quantitative geomorphological analysis, see Ehlen J, Whis-

onant RC, 2008, "Military Geology of Antietam Battlefield, Maryland, U.S.—Geology, Terrain, and Casualties," *Geology Today*, v. 24, n. 1, pp. 20–27.

CHAPTER 3. Killing at Range

1. William Hallahan provides an interesting history of the poor adoption practices of the U.S. military when it comes to rifles in *Misfire: The Story of How America's Small Arms Have Failed Our Military* (1994, New York: Scribner, 580p.)

2. Weaver JR, 2018, *A Legacy in Brick and Stone: American Coastal Defense Forts of the Third System, 1816–1867* (McLean, Va.: Redoubt), 329p.

3. Coggins J, 1990, *Arms and Equipment of the Civil War* (Wilmington, N.C.: Broadfoot), 160p.

4. Or something around a 6-pound solid shot—compositions of shot and shells varied by manufacturer.

5. Early in the war the Confederate armies used six-gun batteries as well.

6. This gun was named after the French emperor Louis Napoleon (Napoleon III), who backed its development.

7. In the winter of 1862 Lee asked for more 12-pounder Napoleons, suggesting melting down all 6-pounder guns and 12-pounder howitzers to provide metal for recasting.

8. Ten- and twenty-pounder varieties were commonly used in the field.

9. Katcher PK, Bryan T, 2001a, *American Civil War Artillery 1861–1865 (1) Field Artillery* (Oxford: Osprey), 48p.

10. Although the captured weapons were not always the highest quality—many forts had artillery pieces that were more than twenty-five years old.

11. Katcher P, 2001b, *American Civil War Artillery 1861–1865 (2): Heavy Artillery* (Oxford: Osprey), 48p.

12. Weaver, 2018.

13. Katcher P, 2001c, *American Civil War Artillery 1861–1865: Field and Heavy Artillery* (Oxford: Osprey Publishing), 100p., p. 62.

14. Abbot HL, 1867, *Siege Artillery in the Campaigns against Richmond, with Notes on the 15-inch Gun, Including an Algebraic Analysis of the Trajectory of a Shot in Its Ricochets upon Smooth Water* (Washington, D.C.: Government Printing Office).

15. Nevertheless, a hit by a mortar shell on a naval vessel was especially devastating because of the paucity of deck armor on most ships.

16. This was especially common on the Mississippi River delta, where Federal shells consistently failed to explode above the surface of the mud.

17. Katcher, 2001c, p. 71.

18. Burns R, 1848, *Questions and Answers on Artillery* (Oxford: Oxford University Press), 110p., p. 11.

19. Lippitt FJ, 1865, *A Treatise on the Tactical Use of the Three Arms: Infantry, Artillery, and Cavalry* (New York: Van Nostrand), 134p., quotation on p. 62.

20. Ehlen J, Whisonant RC, 2008, "Military Geology of Antietam Battlefield,

Maryland, U.S.—Geology, Terrain, and Casualties," *Geology Today*, v. 24, n. 1, pp. 20–27; see also Hippensteel SP, 2016, "Carbonate Rocks and American Civil War Infantry Tactics," *Geosphere*, v. 12, pp. 234–365.

21. The Battle of Antietam occurred in September, so this diminishing effect would probably have been higher if the fighting took place in the late spring when crops were planted.

22. Ehlen and Whisonant, 2008.

23. Guelzo AL, 2013, *Gettysburg: The Last Invasion* (New York: Alfred A. Knopf), 632p., p. 124.

24. Cemetery Hill and Ridge is underlain by a diabase sill. Seminary Ridge is underlain by a dike of similar composition: the dike is a narrower intrusion not quite as suitable for artillery.

25. During the Civil War fuses for all types of shells were notoriously unreliable, especially for Confederate gunners.

26. For some guns more than a dozen types of shell or shot were produced by different manufacturers.

27. The *Official Records* contain several instances of officers praising the weapon's efficacy in combat, and Robert E. Lee greatly desired an increase in the number of these guns for his Army of Northern Virginia. For example, Brig. Gen. and Chief of Ordnance George Ramsay reported that he did not know of a single example of a Napoleon bursting during the first four years of the war, and visiting Austrian army officer FitzGerald Ross stated, "Almost all consider the Napoleon most serviceable" (Katcher and Bryan, 2001a).

28. Griffith P, 1987, *Battle Tactics of the Civil War* (Ramsbury, U.K.: Crowood), 239p.

CHAPTER 4. Geology and Protection

1. Correspondence de Napoléon ler, XIV, no. 12111 (Fuller JFC, 1961, *The Conduct of War, 1789–1961: A Study of the French, Industrial, and Russian Revolutions on War and its Conduct*, London: Eyre & Spottiswoode, p. 52).

2. A distant ancestor of mine, William Tritt, is a good example of a Civil War citizen/soldier. In late August 1862 he was working as a carpenter outside Newville, Pennsylvania. Three weeks later he was firing a Springfield rifle into the Bloody Lane at Antietam.

3. Clausewitz CV, Howard M, Paret P, Brodie B, 1984, *On War* (Princeton, N.J.: Princeton University Press), 732p.

4. Selby JM, 1968, *Stonewall Jackson as Military Commander* (London: Batsford, 251p.), includes a detailed note discussing the origins of Jackson's sobriquet.

5. Carman EA, 2008, *The Maryland Campaign of September 1862: Ezra A. Carman's Definitive Study of the Union and Confederate Armies at Antietam*, ed. Pierro J (London: Routledge), 516p.

6. D. H. Mahan was the father of the famous admiral Alfred Thayer Mahan.

7. Mahan DH, 1856, *A Treatise on Field Fortification, Containing Instructions on the Methods of Laying Out, Constructing, Defending, and Attacking Intrenchments, with the General Outlines Also of the Arrangement, the Attack and Defense of Permanent Fortifications* (New York: J. Wiley), 168p.

8. Mahan DH, 1852, *A Treatise on Field Fortification, Containing Introductions on the Methods of Laying Out, Constructing, Defending, and Attacking Intrenchments, with the General Outlines Also of the Arrangement, the Attack, and Defense of Permanent Fortifications* (New York: J. Wiley), p. 2.

9. An enhanced firing position does not necessarily need to be elevated to an especially high degree; even a four-foot-tall parapet improves a position because it is easier to reload a single-shot musket when standing behind a wall and very difficult when prone.

10. Mahan, 1852, p. 66.

11. Mahan DH, 1865, *An Elementary Course of Military Engineering*, Part 1 (New York: J. Wiley), 284p.

12. Compare images of fresh Civil War parapets and ditches with surviving, unrepaired, or unrestored examples today. Often surviving entrenchments are weathered to the point where only a slight change in relief is evident between the ditch and parapet.

13. Technically, a type of mass wasting described as lateral spread.

14. Unless standing water, insects, and disease became a problem.

15. In complete contrast, engineers were instructed to "sink frequent pits through the clay in order to facilitate its washing" (punch holes through the clay strata to allow water to drain into lower, more permeable layers) on the Mississippi River floodplain. Correspondence between Lieut. Col. Walter B. Scates and Gen. William T. Sherman, 1863, *The War of the Rebellion: A Compilation of the Official Records of the Union and Confederate Armies*, ser. 1, v. 24, pt. 3, quotation on p. 9.

16. Mahan, 1852, p. 35.

17. Mahan DH, 1870, *Field Fortification, Military Mining, and Siege Operations* (New York: J. Wiley), 284p.

18. For a detailed and illustrated discussion of the planning of a proper parapet, see Craig Swain's interesting blog *To the Sound of the Guns*, https://markerhunter.wordpress.com/.

19. Mahan, 1852.

20. Maj. T. B. Brooks, in an engineering note from the *Official Records*, reports on this phenomenon: "Standing between the fires, and within a few yards of the point of striking, the opportunity to observe the effect, in the sand, of these huge shells from the smooth-bore guns of the navy and the rifles of the army was perfect. The ricochet of the former was uniform, and landed nearly every one in the fort. That of the latter was irregular; most of them exploded when they struck throwing up a great quantity of sand, which falls back in its place; hence inflicting no injury save what may come from the heavy jar." *The War of the Rebellion: A*

Compilation of the Official Records of the Union and Confederate Armies, ser. 1, v. 28, pt. I: Reports, quotation on p. 301.

21. Beyond the angle of repose the parapet face becomes an "oversteepened slope."

22. Consider, for example, building a sandcastle at the beach: sandcastles are always easiest to build when the sand is slightly wet but not saturated.

23. Roundness and sphericity have far different meanings to sedimentologists. Spherical grains are round, but round grains are not necessarily spherical—like a kidney bean. An octagon is relatively spherical but far from round.

24. Typically sod if available.

25. This is especially the case if the fortification lacks traverses or the enemy was in a near-enfilade position.

26. This would not be as much of a problem for taller soldiers, but these men would also be more poorly protected by the superior slope.

27. Mahan published multiple important books on field fortifications, including Mahan DH, 1836, *A Complete Treatise on Field Fortification, with the General Outlines of the Principles Regulating the Arrangement, the Attack, and the Defense of Permanent Works* (New York: Wiley & Long); Mahan DH, 1861, *A Treatise on Field Fortification, Containing Instructions on the Method of Laying Out, Constructing, Defending, and Attacking Intrenchments, with the General Outlines Also of the Arrangement, the Attack, and Defense of Permanent Fortifications*, 3rd ed., revised and enlarged (New York: J. Wiley).

28. Along the southeastern U.S. coastline this material often consisted of logs from palmetto trees.

29. Today this same application of coastal vegetation being used to stop the movement of sediment can be observed when new artificial dunes are covered and stabilized with seagrasses.

30. Mahan, 1852, p. 36.

31. These sod blocks also brought some clay to the sandy face of the parapet, slightly enhancing cohesiveness and stability of the underlying sand.

32. Mudstones and shales are both mostly made of clay. Shale is layered, while mudstones are massive (lacking structure like layering or bedding planes).

33. This book primarily focuses on the influence of soil on military engineering and logistics. An entirely different influence of soil and soil types is discussed by John Majewski, who suggests that the nutrient-deprived ultisols of the South were a great disadvantage to the Confederacy, while the less-acidic alfisols of many Northern states were an agricultural advantage. For further discussion of the environmental impact of soils on the war, see Majewski J, 2011, *Modernizing a Slave Economy: The Economic Vision of the Confederate Nation* (Chapel Hill: University of North Carolina Press); Smith AF, 2011, *Starving the South: How the North Won the Civil War* (New York: St. Martin's Press).

34. Rogers JJ, 2015, *A Southern Writer and the Civil War: The Confederate Imagination of William Gilmore Simms* (Lanham, Md.: Lexington), 222p.

35. The regolith is all the material found between the surface of the earth and the hard bedrock below. The soil is the portion of the regolith that supports the growth of plants.

36. Hess EJ, 2005, *Field Armies and Fortification of the Civil War* (Chapel Hill: University of North Carolina Press), p. 147.

37. Hess (2005) noted that Brig. Gen. Alexander Webb's brigade of the Second Corps had no entrenching tools. It should also be pointed out that Cemetery Hill is named after the cemetery located there; the common phrase "six feet under" should give pause to anyone insisting that *no* digging was possible on Cemetery Ridge. Many Union soldiers chose not dig to out of respect for the graveyard, not the geology.

38. This allowed the moat of the fort to be tidal, filling with water every high tide.

39. D. H. Mahan (1870) suggested keeping the sediment moist if at all possible to prevent the fine sand from sifting through revetting materials.

40. Wise SR, 1994, *Gate of Hell: Campaign for Charleston Harbor, 1863* (Columbia: University of South Carolina Press), 312p.

41. Wise, 1994, p. 143.

42. Conchoidal fracture commonly occurs in thick pieces of glass—sharp edges with concentric, curved rings or crescents along the edges. This is why microcrystalline quartz-rich rocks like quartzite, obsidian, or rhyolite were commonly used to make sharp arrowheads or spearpoints.

43. Quartz sand is almost exactly the same hardness (7) as steel (6.5) on the Mohs' scale of hardness.

CHAPTER 5. Durable Rocks and Defensive Stands

1. Zen E, Walker A, 2000, *Rocks and War: Geology and the Civil War Campaign of Second Manassas* (Shippensburg, Pa.: White Mane), 109p. Quartzite is metamorphosed quartz-sandstone.

2. Brown A, 1962, *Geology and the Gettysburg Campaign: Pennsylvania*, Topographic and Geologic Survey Educational Series no. 5, pp. 1–15.

3. Smith RC 2nd, Keen RC, 2008, "Regional Rifts and the Battle of Gettysburg," in Fleeger GM, ed., *Geology of the Gettysburg Mesozoic Basin and Military Geology of the Gettysburg Campaign*, 73rd Annual Field Conference of Pennsylvania Geologists, Gettysburg, Pa., pp. 48–53.

4. Erin Stewart Maudlin (2018) describes another component of erosion that greatly hindered the agricultural productivity of the Confederate states. Southern topsoil eroded quickly, in part because of the climate. Military operations, including clear-cutting of forests, only exacerbated the rate of soil loss. *Unredeemed Land: An Environmental History of Civil War and Emancipation in the Cotton South* (Oxford: Oxford University Press).

5. The ubiquitous phrase "The present is the key to the past" applies to the

concept of uniformitarianism: this application of uniformitarianism is known as "actualism."

6. Mass wasting includes downward movement by soils, rocks, and sediments because of gravity; soil creep, landslides, rock avalanches, and debris flows are all types of mass wasting.

7. These same types of rounded boulders can also be observed across the western portion of the Manassas battlefield in Virginia.

8. Both of these extremely hard-rock battlefields witnessed stones being used as weapons.

9. These karst landscape features are common in the Shenandoah and Hagerstown Valleys.

10. Hippensteel SP, 2019, *Rocks and Rifles: The Influence of Geology on Combat and Tactics during the American Civil War* (Cham, Switzerland: Springer), 321p. (specifically chapter 10).

11. Cuffey RJ, Inners JD, Fleeger GM, Smith RC 2nd, Neubaum JC, Keen RC, Butts L, Delano HL, Neubaum VA, Howe RH, 2006, "Geology of the Gettysburg Battlefield: How Mesozoic Events and Processes Impacted American History," in Pazzaglia FJ, ed., *Excursions in Geology and History: Field Trips in the Middle Atlantic States: Geological Society of America Field Guide*, v. 8, p. 1–16.

CHAPTER 6. Killer Carbonates

1. Formed in calcite-saturated waters in a manner similar to tiny stony hailstones.

2. Sponge spicules are tiny rods and rays of silica or calcite that provide the creature with a degree of structure. Diatom "skeletons" or tests are made of silica and are called frustules.

3. Massive is an understatement: the fort was built from more than 16 million bricks imported from the mainland.

4. Wilson CW Jr, 1964, *Geologic Map and Mineral Resources Summary of the Walterhill Quadrangle*, Tennessee (scale 1:24,000), Geologic Quadrangle Map 315 NW, Tennessee Division of Geology; Wilson CW Jr, 1965, *Geologic Map and Mineral Resources Summary of the Murfreesboro Quadrangle*, Tennessee (scale 1:24,000), Geologic Quadrangle Map 315 SW, Tennessee Division of Geology.

5. The relationship between "killer" limestones like the Conococheague and increased casualty figures was first discussed by Ehlen J, Whisonant RC, 2008, "Military Geology of Antietam Battlefield, Maryland, U.S.—Geology, Terrain, and Casualties," *Geology Today*, v. 24, no. 1, pp. 20–27.

6. Thornberry-Ehrlich TL, 2012, *Stones River National Battlefield: Geologic Resources Inventory Report*, Natural Resource Report NPS/NRSS/GRD/NRR—2012/566 (Fort Collins: National Park Service).

7. Ehlen and Whisonant, 2008; see also Hippensteel SP, 2016, "Carbonate Rocks and American Civil War Infantry Tactics," *Geosphere*, v. 12, pp. 234–365.

8. Stephen Sears (1993) also points out that there were some "limestone ledges that dotted the fields and woods" at Antietam. These offered a small degree of protection, mostly for Confederates. *Landscape Turned Red: The Battle of Antietam* (Boston: Ticknor & Fields), p. 182.

9. Lewis J, 2012, "The Battle of Stones River," *Blue and Gray Magazine*, v. 28.

10. Hunter A, 1904, *Johnny Reb and Billy Yank* (Perrysburg, Ohio: Neal), 720p.

11. See Hess EJ, 2016, *The Rifle Musket in Civil War Combat: Reality and Myth* (Lawrence: University Press of Kansas), 296p.; Nosworthy B, 2003, *The Bloody Crucible of Courage: Fighting Methods and Combat Experience of the Civil War* (New York: Carroll & Graf), 752p., and Griffith P, 2014, *Battle Tactics of the Civil War* (Ramsbury, U.K.: Crowood Press), 240p., for interesting discussions and debates concerning the effective range of rifle muskets in combat and the nature of the employment of Springfield and Enfield rifles during the war.

12. Henderson SW, 2004, "The Geology of the Chickamauga Campaign, American Civil War," in Caldwell DR, Ehlen J, Harmon RS, eds., *Studies in Military Geography and Geology* (New York: Springer, Dordrecht), pp. 173–84.

13. From a geological perspective, it is a shame George "The Rock of Chickamauga" Thomas never met Thomas "Stonewall" Jackson in battle.

14. Henderson, 2004.

15. Nosworthy, 2003.

CHAPTER 7. Battling in the Basins

1. Hard rocks usually outcrop in higher proportions compared with softer rocks, thus they are more visible on the battlefield.

2. Thornberry-Ehrlich T, 2008, *Manassas National Battlefield Park Geologic Resource Evaluation Report*, Natural Resource Report, NPS/NRPC/GRD/NRR—2008/050 (Denver: Geologic Resources Division).

3. Diabase is very similar to the more common diorite—an intermediate igneous rock that looks like a salt-and-pepper form of dark granite (more pepper).

4. Toewe EG, 1966, *Geology of the Leesburg Quadrangle, Virginia*, Virginia Division of Mineral Resources Report of Investigation, n. 11, 52p.

5. Smith RC 2nd, Keen RC, 2008, "Regional Rifts and the Battle of Gettysburg," in Fleeger GM, ed., *Geology of the Gettysburg Mesozoic Basin and Military Geology of the Gettysburg Campaign*, 73rd Annual Field Conference of Pennsylvania Geologists, Gettysburg, Pa., pp. 48–53. Dikes are igneous intrusions that cut across the preexisting strata, while sills are concordant and extend along the preexisting bedding planes. Seminary Ridge is a dike, while Cemetery Ridge was formed by a sill.

6. National Park Service display on Lee's Hill ("The Pioneers"), Fredericksburg National Battlefield.

CHAPTER 8. Sedimentary Geology and Logistics

1. Historian Earl J. Hess has produced a volume discussing military logistics that focuses less on geology and more on organization and coordination. See Hess EJ, 2017, *Civil War Logistics: A Study of Military Transportation* (Baton Rouge: Louisiana State University Press), 341p.

2. "Fall Line" seems appropriate for the location on the rivers where the contact can clearly be delineated. For areas covered by soil the contact may be buried and obscured, so "Fall Zone" is more appropriately broad. References to the Fall Zone can be found in the *Official Records*. For example, Maj. Gen. John W. Geary of the XX Corps was certainly talking about crossing the Fall Zone in South Carolina in this passage from his report: "Across the Saluda the geological features of the country present a sudden change; red and yellow clay and silex predominate, and after crossing Broad River granite bowlders [sic] are abundant." Note: "silex" is a term for unconsolidated sediments. *The War of the Rebellion: A Compilation of the Official Records of the Union and Confederate Armies*, ser. 1, v. 47, pt. 1: Reports, quotation on p. 718.

3. Kilpatrick D, 2007, *Logistics of the American Civil War*, RUSI Journal, v. 152, pp. 76–81, esp. p. 77.

4. Hess, 2017.

5. Hess, 2017, subsection "Railroad Management under Halleck and Grant."

6. The most comprehensive study of the challenges of supplying a large army campaign was recently completed by Earl J. Hess (2020). He details in particular the logistical nightmare Sherman and Grant faced when invading the Deep South. *Civil War Supply and Strategy: Feeding Men and Moving Armies* (Baton Rouge: Louisiana State University Press), 448p.

7. Lackey RC, Notes on Civil War Logistics: Facts & Stories (https://transportation.army.mil/History/PDF/Peninsula%20Campaign/Rodney%20Lackey%20Article_1.pdf; accessed November 10, 2020).

8. Coggins J, 1990, *Arms and Equipment of the Civil War* (Wilmington, N.C.: Broadfoot), p. 121.

9. Garrison W, 2001, *Webb Garrison's Civil War Dictionary* (Nashville: Cumberland House), 350p.

10. Magner GE, 1963, *Porosity and Bulk Density of Sedimentary Rocks*, United States Geological Survey Bulletin 1144-E, 60p.

11. Gravel roads suffered less deterioration from the weather than did unpaved roads, but they still were not impervious, so to speak, to the effects of thousands of boots and hundreds of wagon or caisson wheels. By the end of a campaign macadam roads were often in need of maintenance or repair.

12. The longest plank road in the United States traversed the 120 miles between Fayetteville and Winston-Salem (*Official Records of the War of the Rebellion*, v. 47, p. 1,084).

13. A more detailed discussion of the construction of corduroy roads can be found in *Ruin Nation: Destruction and the American Civil War* by Megan Kate Nel-

son (2012; University of Georgia Press). The author also analyzes the conse-quences of building such transportation routes from an environmental perspec-tive—a vicious cycle where the timber cutting that was necessary to provide the planks for the roads also led to increased soil erosion and runoff, and thus the need for more corduroy roads.

Although corduroy roads could be constructed rather rapidly when appropri-ate material was available, it was often an arduous task. Sherman's greatest engi-neer, Col. Orlando Poe, remarked that constructing corduroy roads "was a very simple affair where there were many fence-rails, but, in their absence, involved the severest of labor." Nichols GW, 1865, *The Story of the Great March from the Di-ary of a Staff Officer* (New York: Harper).

14. When pulled around a curve, railcars will try to follow the chord of the curve and not the arc.

CHAPTER 9. A River Runs through It

1. The general orientation of the Potomac, Rappahannock, York, and James Rivers can be linked to the massive underlying impact crater created by a me-teorite strike that occurred 35 million years ago during the Eocene epoch at the mouth of the Chesapeake Bay. This 50-mile-wide topographic depression altered the path of the regional rivers and disrupts aquifers even to the present day.

2. There was also certainly a downside to conducting operations around marshy ground and bogs. Corinth, Mississippi, for example, was notorious for having terrible drinking water and ferocious biting insects.

3. Water that was waist-deep or greater might also saturate gunpowder, elimi-nating the effectiveness of the primary weapon the wading soldier was carrying.

4. The irregular shape of the mid-Atlantic coastline, and thus the shape of the peninsula, are the result of the Atlantic Ocean transgressing across the rela-tively flat Coastal Plain. The creation of this terrain, and the further influence of weather and climate on McClellan's campaign, are discussed in chapter 6 of Har-old Winters's *Battling the Elements: Weather and Terrain in the Conduct of War* (1998; John Hopkins University Press).

5. The effects of mud on straggling and morale during the Peninsula Cam-paign are documented in more detail in Shively Meir K, 2013, *Nature's Civil War: Common Soldiers and the Environment in 1862 Virginia* (Chapel Hill: University of North Carolina Press).

6. Depending on the location of the river, this downcutting eroded into por-tions of the Bacons Castle Formation (Upper Pliocene, sands and gravels) or Chesapeake Group (Miocene, clay, silt, and sand). There are also stretches of the river that incise into Pliocene sands and gravels.

7. Berquist CR Jr, 2012, "Mapping Bottom Sediments in the James and Chickahominy River Estuaries, Virginia," *Geological Society of America Abstracts with Programs*, v. 44, p. 21.

8. The river discharge (volume/time or cubic meters per second) may increase

by a factor of five, but the elevation of the water may only increase by a factor of less than one. When a river floods onto a broad floodplain, vast volumes of water will be spread across the plain with little rise in stage (elevation).

9. Bailey RH, 1983, *Forward to Richmond* (New York: Time-Life), 176p.

10. River velocity and erosional power are directly related to discharge: flood waters carry more sediment and have more erosional potential, especially for infrastructure.

11. Brent Nosworthy (2003) describes the options available for crossing the river to the Fifty-Seventh New York during the early stages of the Battle of Fair Oaks. After observing the poor conditions and deterioration of the only bridge available along their route, the men elected to wade across the river. They found the water only waist-deep, so the men lifted their rifles and ammunition to protect them from getting wet and began to splash across the half-mile-wide floodwaters. After crossing the river, the men found that the sand and silt from the newly disturbed riverbed had made its way into their shoes, where it proved difficult to dislodge and a severe irritant to their already sore feet. Nosworthy B, 2003, *The Bloody Crucible of Courage: Fighting Methods and Combat Experience of the Civil War* (New York: Carroll & Graf), 752p.

12. Broadwater RP, 2014, *The Battle of Fair Oaks: Turning Point of McClellan's Peninsula Campaign* (Jefferson, N.C.: McFarland), 219p. See also Bailey, 1983.

13. Sears SW, 1992, *To the Gates of Richmond: The Peninsula Campaign* (Boston: Ticknor & Fields), 468p.

14. Gaines' Mill battleground is underlain by the Chesapeake Group to the east and Pliocene sand and gravel to the west.

15. Interesting note: Only eight U.S. cities in 1862 had a population larger than McClellan's army of 120,000 men, and seven of these municipalities were in the North.

16. Papanek JL, ed., 1984, *Lee Takes Command: From Seven Days to Second Bull Run* (New York: Time-Life), 176p. Jackson's lethargy is discussed on pp. 54–55.

17. Sears, 1992, p. 288.

18. Most of the Malvern Hill battlefield is underlain by the Bacons Castle Formation, although the low-lying area to the east, beyond the bluffs of the western side of the hill, is covered by the much younger river terraces of the Quaternary Chuckatuck Formation.

19. *Official Records*, ser. 1, v. 11, no. 2, p. 492.

20. *Official Records*, ser. 1, v. 11, no. 2, p. 497.

21. Harold Winters (1998) discusses how Antietam Creek became so difficult to cross, despite not being particularly deep: "Because the Potomac is a powerful river, it has cut downward into the low, rolling hills immediately west of the Blue Ridge. As a result, the entry level of Antietam Creek has also been lowered. The lower confluence increases the creek's gradient, velocity, erosive power, and transport capabilities, allowing it to form a shallow yet steep-walled valley." *Battling the Elements: Weather and Terrain in the Conduct of War* (Baltimore: John Hopkins University Press), quotation on p. 125.

22. Mixon RB, Pavlides L, Powars DS, Froelich AJ, Weems RE, Schindler JS, Newell WL, Edwards LE, Ward LW, 2000, *Geologic Map of the Fredericksburg 30′ x 60′ Quadrangle, Virginia and Maryland* (Scale 1:100,000), USGS, I-2607.

23. Murray W, 2016, *A Savage War: A Military History of the Civil War* (Princeton, N.J.: Princeton University Press), 616p.

CHAPTER 10. Burnside and the Bluffs

1. White JE, 2018, *New Bern and the Civil War* (Charleston, S.C.: History Press), 208p.

2. See Virginia Department of Mines, Minerals, and Energy's interactive map of surficial geology, https://www.dmme.virginia.gov/webmaps/DGMR/ (accessed January 3, 2021).

3. Perhaps a better term for these "snipers" would be "sharpshooters." The sniper role, so well established during the world wars, really did not exist in the Civil War.

4. Ehlen J, Abrahart RJ, 2002, "Effective Use of Terrain in the American Civil War: The Battle of Fredericksburg, December 1862," in Doyle P, Bennett MR, eds., *Fields of Battle*, GeoJournal Library, v. 64 (New York: Springer, Dordrecht), pp. 63–97.

5. Ehlen and Abrahart, 2002.

6. George C. Rable discusses the "digging-in" of Lee's army in *Fredericksburg! Fredericksburg!* (2002; University of North Carolina Press).

7. Chapter 11 details the debacle of Burnside's January 1863 maneuvering in the mud.

CHAPTER 11. Sediments and Morale

1. Kehew AE, 1998, *Geology for Engineers and Environmental Scientists* (Hoboken, N.J.: Prentice Hall), 574p.

2. U.S. Marine Corps, *Small Unit Leaders Guide to Weather and Terrain*, MCRP 12–10.1, 149p.

3. Smith GW, 2010, *Battles and Leaders of the Civil War, v. 2: The Struggle Intensifies. Two Days of Battle at Seven Pines (Fair Oaks)* (Auckland, N.Z.: Castle), 750p.

4. Morgan WH, 1911, *Personal Reminiscences of the War of 1861–1865* (Lynchburg, Va.: J. P. Bell), 264p.

5. Rhodes EH, 2010, *All for the Union: The Civil War Diary and Letters of Elisha Hunt Rhodes* (New York: Knopf Doubleday), 272p., quotation on p. 46.

6. This occurred during the Battle of Secessionville (or First Battle of James Island) on June 16, 1862.

7. Wood CE, 2006, *Mud: A Military History* (Lincoln, Nebr.: Potomac), 190p.

8. For example, S. G. Van Anda, the lieutenant colonel of the Twenty-First Iowa, described how sediments of all forms slowed his march during the Mobile Campaign. "We found the roads dry, but marching heavy on account of the

sand," he complained in a report to his superior. Once off sands of the barrier islands, the "country became of a very difficult character for transportation and artillery, being exceedingly wet and marshy." *Official Records of the Union and Confederate Armies,* ser. 1, v. 49, p. 168.

9. *Official Records of the Union and Confederate Armies,* ser. 1, v. 49, p. 10.

10. Alexander Downing of the Eleventh Iowa was describing "churnability" in this diary passage: "The landing place is nothing but a jelly of mud—there are so many mules, horses and men passing back and forth." Downing A, 1916, *Downing's Civil War Diary* (Des Moines: Historical Department of Iowa).

11. Grant US, 1903, *The Personal Memoirs of U.S. Grant* (New York: Century), 584p.

12. When mud is sticking to other items, it is being adhesive; when sticking to itself, cohesive. Both properties act together to pull off a soldier's boots when marching on a really muddy road.

13. Letter from Crozer to Oatman, March 11, 1863 (Crozer Papers, United States Military Academy Library, West Point, New York).

14. Rahn PH, 1996, *Engineering Geology: An Environmental Approach* (Hoboken, N.J.: Prentice Hall), 657p.

15. Prothero DR, Schwab F, 2013, *Sedimentary Geology* (New York: W. H. Freeman), 593p.

16. Environmental Protection Agency, 1995, *Compilation of Air Pollution Emission Factors,* v. 1: *Stationary Point and Area Sources,* 29p.

17. An example of a point source for dust during the Civil War would be the cannon muzzle and associated black-powder smoke from an artillery piece.

18. Scribner BF., 1887, *How Soldiers Were Made: or, The War as I Saw It under Buell, Rosecrans, Thomas, Grant and Sherman* (Chicago: Donohue & Henneberry), quotation on p. 56.

19. Jaquette HS, ed., 1971, *South after Gettysburg: Letters of Cornelia Hancock from the Army of the Potomac, 1863–1865* (Freeport, N.Y.: Books for Libraries), quotation on pp. 114–15.

20. Catton B, 1953, *A Stillness at Appomattox* (New York: Anchor), 448p.

21. Downing, 1916, p. 131. Downing makes more than twenty-five references to mud and dust in his diary.

22. Holcomb W, 2008, The Mud March N-6, http://fredmarkers.umwblogs. org/2008/03/03/the-mud-march/ (accessed October 26, 2019).

23. Ambrose K, Henry D, Weiss A, 2002, *Washington Weather* (Gallatin, Tenn.: Historical Enterprises), 252p.

24. Winters H, 1991, "The Battle That Was Never Fought: Weather and the Union Mud-March of January 1863," *Southeastern Geographer,* v. 31, pp. 31–38.

25. De Loss Love, W, 1866, *Wisconsin in the War of the Rebellion: A History of All Regiments and Batteries the State Has Sent to the Field, and Deeds of Her Citizens, Governors and Other Military Officers, and State and National Legislators to Suppress the Rebellion, Part 1* (Chicago: Church & Goodman), 1,140p.

26. Rhodes, 2010, p. 89.

27. Williams AS, 2015, *From The Cannon's Mouth: The Civil War Letters of General Alpheus S. Williams* (Auckland, N.Z.: Pickle Partners), 349p.

28. Construction of corduroy roads usually involved great deals of deforestation and, as Joan Cashin (2018) states, "Deforestation creates the perfect conditions for oceans of mud." The result, in short, was a vicious mud and wood cycle. *War Stuff: The Struggle for Human and Environmental Resources in the American Civil War* (Cambridge: Cambridge University Press), quotation on p. 104.

29. Swartz B, 2014, http://maineatwar.bangordailynews.com/2014/12/17/mud-on-the-mules/ (accessed November 3, 2019).

30. Finfrock B, 2013, *Across the Rappahannock* (Berwyn Heights, Md.: Heritage), 280p.

31. Lemuel Abijah Abbott of the Tenth Vermont wrote about the difficulty of traversing across—and through—sediment in southern Virginia: "marching through sand ankle deep as dry as an ash heap with the air so thick with dust one a few steps away is invisible" (Personal Recollections and Civil War Diary, 1864 / 1908, Harvard University).

32. Rhodes, 2010, p. 125.

33. Winschel TJ, 2001, *The Civil War Diary of a Common Soldier: William Wiley of the 77th Illinois Infantry* (Baton Rouge: Louisiana State University Press), 216p.

34. Journal entry from Carlos W. Colby, excerpted in *American Civil War: Interpreting Conflict through Primary Documents* (v. 1, 2019, Justin D. Murphy, ed.; ABC-CLIO).

35. The Homestead Act, which was enacted in 1862, allowed American citizens to claim 160 acres of surveyed government land as long as they agreed to improve the plot with a dwelling and to farm the land.

36. Campbell JQA, 2000, *The Union Must Stand: The Civil War Diary of John Quincy Adams Campbell, Fifth Iowa Volunteer Infantry* (Knoxville: University of Tennessee Press), 267p.

CHAPTER 12. Geology of the Father of Waters

1. Smith S, 2019, The River War, https://www.battlefields.org/learn/articles/river-war (accessed July 23, 2019).

2. Korn J, 1985, *War on the Mississippi: Grant's Vicksburg Campaign* (New York: Time-Life), 176p.

3. Fort Henry was the easier of the two targets to capture because of poor military engineering: the fort was constructed on the floodplain of the river, not above it (as with Fort Donelson). At the time of Grant's assault, a significant portion of Fort Henry's firepower was, quite figuratively, under water. Browning J, Silver S, 2020, *An Environmental History of the Civil War* (Chapel Hill: University of North Carolina Press), 272p.

4. When the later supercontintent Pangea rifted in the Triassic, the regions

around Gettysburg, Pennsylvania, and Manassas, Virginia, were in grabens. These valleys became lakes and sediment traps, producing the distinct red sandstones and shales that underlie the battlegrounds.

5. Galicki S, Schmitz D, 2016, *Roadside Geology of Mississippi* (Missoula, Mont.: Mountain Press), 288p., quotation on p. 13.

6. Galicki and Schmitz, 2016, p. 20.

7. Ward LW, Powars DS, 2004, "Tertiary Stratigraphy and Paleontology, Chesapeake Bay Region, Virginia and Maryland, July 15–July 17 (Field Trip Guidebooks)," chapter 9 in Eastabrook J, Schindler K, eds., *Tertiary Lithology and Paleontology, Chesapeake Bay Region* (PDF), USGS Circular (Report), 1264 (Eastern Region Publications, Geologic Discipline, United States Geological Survey).

8. Hippensteel SP, 2019, *Rocks and Rifles: The Influence of Geology on Combat and Tactics during the American Civil War* (Cham, Switzerland: Springer), 321p., chaps. 9 and 11; Cross A, Berquist CR, Occhi M, Strand J, Bailey CM, 2017, "Geology and the Petersburg Campaign in the American Civil War," in Bailey CM, Jaye S, eds., *Blue Ridge to the Beach: Geological Field Excursions across Virginia: Geological Society of America Field Guide 47* (Boulder, Colo.: Geological Society of America), pp. 163–74.

9. Gushing EM, Boswell EH, Hosman RL, 1964, *General Geology of the Mississippi Embayment*, Geological Survey Professional Paper 448-B, 32p.

10. Gushing et al., 1964, p. 2.

11. Spearing D, 2007, *Roadside Geology of Louisiana* (Missoula, Mont.: Mountain Press), 225p.

12. Krinitzsky EL, Turnbull WJ, 1967, *Loess Deposits of Mississippi*, v. 94, Geological Society of America Special Paper (Boulder, Colo.: Geological Society of America), 64p.

13. Sea level typically falls during glacial intervals. As the glaciers advanced across the valley from the north, the shoreline regressed toward the south.

14. Powell JW, 1885, *Annual Report of the Director of the United States Geological Survey to the Secretary of the Interior*, v. 6, Geological Survey (U.S.), 570p.

15. Bettis EA 3rd, Muhs DR, Roberts HM, Wintle AG, 2003, "Last Glacial Loess in the Conterminous USA," *Quaternary Science Reviews*, v. 22, pp. 1907–46.

16. Environmental historian Lisa M. Brady (2012) described the dual nature of the levees on the Southern states: "They provided rich, dry land for the region's cotton cultivation, but they also formed the edges of the area's nearly impassable swamps." *War upon the Land: Military Strategy and the Transformation of Southern Landscapes during the American Civil War* (Athens: University of Georgia Press), p. 27.

17. Cohesive clay would represent the only sediment that might be difficult to excavate.

18. Krinitzsky and Turnbull, 1967. Typical mineralogical breakdown of silt: 60–65 percent quartz, 15–20 percent carbonate grains, 6–8 percent clay, 5–7 percent feldspar, and 2–5 percent other minerals.

19. In 1915 R. E. Dodge noted that these cutoffs aren't always the best for navigation, as they are often made in streams with a "continually shifting channel." *Advanced Geography* (Chicago: Rand McNally), p. 32.

20. VanArsdale V, 2014, "The New Madrid Seismic Zone of the Central United States," in Talwani P, ed., *Intraplate Earthquakes* (Cambridge: Cambridge University Press), pp. 162–97.

21. This reversal occurred between Islands Number 10 and 8. Knox R, Stewart D, 1995, *The New Madrid Fault Finders Guide* (Marble Hill, Mo.: Gutenberg Richter), 179p.

22. VanArsdale RB, 2009, *Adventures through Deep Time: The Central Mississippi River Valley and Its Earthquakes*, v. 455, Geological Society of America Special Paper (Boulder, Colo.: Geological Society of America), 107p.

23. Pratt F, 1956, *Civil War on Western Waters* (New York: Henry Holt), 255p. Christopher Gabel (2001) points out the advantages of attacking head-on from downstream: the head-on attack position takes advantage of the ironclad's heaviest armor, and the boats were easier to maneuver when facing into the current. *Staff Ride Handbook for the Vicksburg Campaign, December 1862–July 1863* (Fort Leavenworth, Kans.: Combat Studies Institute, U.S. Army Command and General Staff College), 240p.

24. Brady, 2012, 189p.

25. Army TF Jr, 2016, *Engineering Victory: How Technology Won the Civil War* (Baltimore: Johns Hopkins University Press), 369p.

CHAPTER 13. The Vicksburg Campaign

1. The fine white sand of Ship Island was of great interest to the Union soldiers who captured the island. "Virtually all Union soldiers who wrote about Ship Island commented on the sand," noted David Rankin in *Diary of a Christian Soldier: Rufus Kinsley and the Civil War* (2004, Cambridge University Press), n25.

2. According to the United States Army Corps of Engineers (Engineer Research and Development Center), despite local subsidence and sea-level rise, Pass-a-Loutre has been getting shallower in the last fifty years due to increased sediment deposition. Little CD Jr, 2014, *Geomorphic Assessment of Pass a Loutre and South Pass, Mississippi River Delta,* U.S. Army Engineer Research and Development Center, Coastal and Hydraulics Laboratory.

3. The *Colorado* had forty-four cannon. Nineteen guns were removed from the ship, likely in an attempt to lighten her and decrease the amount of water she drew. When the ship still could not pass over the bar, these guns were redistributed among the other ships headed upriver.

4. Greene J, 1982, *Special History Study: The Defense of New Orleans, 1718–1900* (Washington, D.C.: National Park Service), 614p.

5. Pratt F, 1956, *Civil War on Western Waters* (New York: Henry Holt), p. 33.

6. Chaitin PM, 1984, *The Coastal War: Chesapeake Bay to Rio Grande* (New York: Time-Life), 176p., use of sediment described on p. 66.

7. Korn J, 1985, *War on the Mississippi: Grant's Vicksburg Campaign* (New York: Time-Life), p. 19.

8. William Shea and Terrence Winshel (2005) detail the military challenges presented to both sides of the conflict by the topography of the Mississippi River Valley. *Vicksburg Is the Key: The Struggle for the Mississippi River* (Lincoln, Nebr.: Bison Books).

9. Gabel CR, 2001, *Staff Ride Handbook for the Vicksburg Campaign, December 1862–July 1863* (Fort Leavenworth, Kans.: Combat Studies Institute, U.S. Army Command and General Staff College), 240p.

10. Williams T, June 6, 1862, "Letters" to Mary Neosho Williams, American Historical Review, v. 14, no. 2, pp. 304–28, quotation on p. 322.

11. Even if the new channel was not entirely out of range of the closest Confederate batteries, they would make a more difficult, long-range, moving target.

12. Mason FH, 1876, *The Forty-Second Ohio Infantry: A History* (Cleveland: Cobb, Andrews), 328p.

13. Grant also schemed to cut a canal to Walnut Bayou for the lightest of his ships. This was not a major engineering project designed to bypass the guns of Vicksburg, only a way to provide supplies and additional troops to his army south of the city. As with several other projects, the "Duckport Canal" was abandoned when the river level began to fall.

14. Schmelzer PL, 2013, "Clausewitz on the Operational Art as Practiced in the Vicksburg Campaign," in Woodworth SE, Grear CD, eds., *The Vicksburg Campaign*, March 29–May 18, 1863 (Carbondale: Southern Illinois University Press), 239p.

15. Army TF Jr, 2016, *Engineering Victory: How Technology Won the Civil War* (Baltimore: Johns Hopkins University Press), pp. 180–81.

16. Donald L. Miller's *Vicksburg: The Campaign That Broke the Confederacy* (2019; Simon & Schuster) discusses the efforts of the Federal sappers to dig through enough clay that they might reach a layer of sand, after which the erosive power of the Mississippi could be used to further excavate the canal project.

17. Harrelson DW, Zakikhani M, Tillotson AL, 2016, "Canals, Cutoffs, and the Vicksburg Campaign," *Geostrata*, pp. 28–32.

18. Army, 2016, p. 181.

19. The combination of sedimentology (the cohesiveness of the clay) and stratigraphy (the layering of the sediments) provided problems for the canal builders. Charles A. Dana, a "special commissioner" of the U.S. War Department, recorded this in his correspondence: "The water is now flowing freely through the whole length of the canal opposite Vicksburg, but produced no effect in wearing away the compact clay soil, which in the lower half of its course is especially tenacious." The clay becomes more difficult to work with as the canal cuts south (downhill), because the trace of the canal is plunging deeper into the underlying clay strata. *The War of the Rebellion: A Compilation of the Official Records of the Union and Confederate Armies*, ser. 1, v. 24, p. 65.

20. Grant US, 1894, *Personal Memoirs of U.S. Grant* (New York: Webster), p. 265.

21. In his personal memoirs Grant acknowledges that the canal, even if successfully completed, could only be used at night. As soon as the Confederates across the river realized the scope of the project, they had constructed a battery to fire lengthwise down nearly the entirety of the canal.

22. McPherson was the superintending engineer during the fortification of Alcatraz Island. When he was killed during the fighting around Atlanta, he became the second highest-ranking Federal officer killed during the war.

23. Harrelson et al., 2016.

24. Army, 2016, p. 174.

25. Hankinson, A, 1993, *Vicksburg 1863: Grant Clears the Mississippi* (Oxford: Osprey), 96p.

26. Fort Pemberton was "constructed of cotton bales covered over with sand and earth, and in itself would be very valuable," according the official report of Lieut. Col. J. H. Wilson. *The War of the Rebellion: A Compilation of the Official Records of the Union and Confederate Armies*, ser. 1, v. 24, quotation on p. 381.

27. Army, 2016, p. 176.

28. Also, approaching ships would only be able to bring their bow-mounted artillery into the battle because of the narrowness of the river along this sector.

29. Army, 2016, p. 177.

30. Hankinson, 1993, p. 32.

31. James McPherson (2003) summarizes the relationship between Grant's soldiers and mud in an artful manner: "For two months Grant's army had been floundering in the mud. Many of them rested permanently below the mud, victims of pneumonia or typhoid or dysentery or any of a dozen other maladies." *Battle Cry of Freedom* (Oxford: Oxford University Press), p. 588.

32. Korn, 1985, p. 85.

33. Gabel, 2001, p. 106.

34. Grabau (2000) discusses the geographic setting. *Ninety-Eight Days: A Geographer's View of the Vicksburg Campaign* (Knoxville: University of Tennessee Press), 680p., p. 19.

35. National Park Service, 2019, https://www.nps.gov/vick/learn/history culture/battleportgibson.htm (accessed October 16, 2019).

36. Bicker AR, 1969, *Geologic Map of Mississippi*, Mississippi Department of Environmental Quality, Office of Geology.

37. Army, 2016, notes than only ~500 shovels were available to be dispersed across 18,500 combatants.

38. Justin Solonick (2015) points out that the field manuals in use at the time were designed for the cohesive soils of the East; parapets made of disturbed loess would need to be thicker to accomplish their designed defensive purpose. *Engineering Victory: The Union Siege of Vicksburg* (Carbondale: Southern Illinois University Press), 304p.

39. Field R, 2007, *American Civil War Fortifications (3): The Mississippi and River* (Oxford: Osprey), 64p.

40. Gabel, 2001, pp. 32–34.

41. Gabel (2001) refers to these ridges and valleys thusly: "a fantastic array of razor-back ridges and precipitous ravines" (p. 71).

42. Solonick, 2015.

43. Grant US, 1894, *Personal Memoirs of U.S. Grant* (New York: Webster), p. 314.

44. Field, 2007, 64p. Grant had an artillery advantage, in terms of the number of guns, but he lacked heavy siege guns until he turned to the navy for help.

45. Hankinson, 1993, p. 139.

46. Bradley Clampitt discusses the civilian exploitation of loess for underground dwellings in *Occupied Vicksburg* (2016; Louisiana State University Press).

47. Grabau, 2000, p. 21.

48. Solonick (2015) is one of only a few texts that discusses the benefits of loess for creating gun emplacements.

49. Grabau, 2000, p. 48.

50. A comparison between the mining operations at the Third Louisiana Redan at Vicksburg and the mine at Petersburg that led to the Battle of the Crater: Vicksburg: shaft = 3' x 4', 40' long; resulting crater 40' wide x 12' deep; ~50 casualties; took 10 days to dig. Petersburg: shaft = 3' x 4.5', 511' long; resulting crater 170' wide x 110' deep; 278 casualties; took more than 30 days to dig. The mine and gallery at Vicksburg proved much easier to construct and required a fraction of the bracing and drainage needed in the sands and clays at Petersburg.

51. Army, 2016, p. 202.

52. Grant, 1894, p. 325.

53. The target was a Yankee sap roller, but a small, poorly sited charge failed to accomplish much.

54. Loess is porous but not particularly permeable. Warren Grabau (2000) points out that this lack of permeability prevented rainwater from providing good, shallow drinking water wells, and of the rare springs, "nearly all go dry in July, August, and September" (p. 21).

55. Hippensteel SP, 2019, *Rocks and Rifles: The Influence of Geology on Combat and Tactics during the American Civil War (Cham, Switzerland: Springer),* 321p., specifically chap. 11—Petersburg.

CHAPTER 14. Protecting the Shoreline

1. Weaver JR, 2018, *A Legacy in Brick and Stone: American Coastal Defense Forts of the Third System, 1816–1867* (McLean, Va.: Redoubt), p. 5.

2. Mahan DH, 1870, *Field Fortifications, Military Mining, and Siege Operations* (New York: J. Wiley), two volumes totaling 650p.

3. Fort Moultrie and Fort Sumter, lying on opposite sides of Charleston Har-

bor, offer an interesting contrast in the use of sand parapets as defensive measures to protect a brick scarp. Fort Moultrie has great masses of sand added to
the exterior of the fort, greatly improving the strength of the fort and allowing it to be used through the end of World War II. Sumter's scarp is surrounded
by water, prohibiting the use of sand. This fort was rendered obsolete after the
Spanish-American War.

4. The vulnerability of fortifications from this time period to modern rifled artillery was demonstrated when the U.S. Navy bombarded the forts of San Juan,
Castillo San Felipe del Morro, and Castillo San Cristóbal on May 12, 1898. Under U.S. control, reinforced concrete was added to modernize these forts for later
conflicts.

5. The current estimate for the length of the U.S. shoreline from the National
Oceanic and Atmospheric Administration is 95,471 miles. Of course, prior to the
Civil War the length of the shoreline did not include Hawaii or Alaska.

6. Weaver, 2018, p. 6.

7. "Coastal erosion" represents the combined threat from beach drift, sea-
level rise, and storm activity.

8. Possibly a bioturbation structure called *Thalassinoides*. This trace fossil exploded in abundance during the Mesozoic and is commonly associated with burrowing creatures similar to crayfish.

Cross-bedding forms when bedding planes form at an angle, giving the cross-
section of the rock the appearance of a herringbone pattern. Cross-bedding is
common in sand dunes and shallow rivers and streams. Planar bedding forms
when the beds are parallel and horizontal.

9. Weaver, 2018, p. 21.

10. This resistance is minimized if the sediment is saturated.

11. Interestingly, the blockade squadron never had to deal with a significant
hurricane during the entirety of the Civil War. Ken Noe (2014) points out that
the 1860s were, for the fortunate Federal navy, one of the quietest decades for
hurricanes in the last 160 years. "Fateful Lightning: The Significance of Weather
and Climate to Civil War History," in Drake BA, ed., *The Blue, the Gray, and the
Green: Toward an Environmental History of the Civil War* (Athens: University of
Georgia Press), pp. 16–33.

12. William Trotter (1989) states that three thousand shells were fired at the
forts at a rate of up to twenty-eight rounds per minute. *Ironclads and Columbiads*
(Winston-Salem, N.C.: John F. Blair), 456p.

13. Dougherty K, 2010, *Strangling the Confederacy: Coastal Operations in the
American Civil War* (Havertown, Pa.: Casemate), 233p.

14. Simons G, 1983, *The Blockade: Runners and Raiders* (New York: Time-Life),
176p.

15. Page D, 1994, *Ships versus Shore: Civil War Engagements along the Southern
Shores and Rivers* (Nashville: Rutledge Hill Press), 410p.

16. In September 1861 the Atlantic Blockade Squadron was subdivided based

on latitude. The South Atlantic Squadron concentrated on the coastal region between Cape Fear, North Carolina, and Georgia.

17. Imagine the frustration of climbing the career ladder to find yourself at the top—a general—only to be known by the men you command as "the other" Sherman.

18. Dougherty, 2010, p. 50.

19. Sand even rendered most of the small arms available to the men worthless. The commander of Fort Walker, Col. John Wagener, noted, "They [his men] had been working at the guns in most cases in shirt sleeves; the sand had covered their knapsacks and muskets, sometimes two or three feet deep, and very few arms were therefore brought off." *The War of the Rebellion: A Compilation of the Official Records of the Union and Confederate Armies*, ser. 1, v. 6, p. 16.

20. Gillmore Q, 1890, *The War of the Rebellion: A Compilation of the Official Records of the Union and Confederate Armies*, ser. 1, v. 28, pt. 1, pp. 37, 228.

21. Moore MA, 1999, *Moore's Historical Guide to the Wilmington Campaign and the Battles for Fort Fisher* (Mason City, Iowa: Savas), 210p.

22. Nevertheless, in most ship-to-shore encounters, the Union naval vessels carried larger-caliber guns.

23. Penetration of the hull could cause flooding, damage boilers and machinery, kill crewmembers, or, most devastatingly, compromise the magazine.

CHAPTER 15. The Education of Quincy Gillmore

1. Young R, 1947, *Robert E. Lee and Fort Pulaski*, Popular Series 11 (Washington, D.C: National Park Service), https://www.nps.gov/parkhistory/online_books/popular/11/index.htm.

2. I have hand-driven hundreds of gouge augers into similar marsh strata up and down the southeastern Atlantic coast to study the barrier island stratigraphy and search for ancient hurricane deposits. Drilling through the mud is as simple as gently pushing the hand auger into the sediment, but you must be careful when striking the Pleistocene "basement" sand—it feels as if you have struck a cement floor with a startling jolt.

3. Fort Pulaski was originally planned to be nearly identical to Fort Sumter in design, but the unstable marsh surface prohibited the addition of a second story and third tier of guns.

4. Lisa M. Brady discusses the challenges that sedimentary geology and geography posed to Sherman's capture of Savannah after his march to the sea. The city sits on a sandy barrier, overlooking the marshes and swamps of the Savannah and Ogeechee Rivers. The Confederates were quick to place heavy artillery to cover any permissible pass through the muddy terrain, limiting Sherman's options for attack. Brady LM, 2012, *War upon the Land: Military Strategy and the Transformation of Southern Landscapes during the American Civil War* (Athens: University of Georgia Press), 179p.

5. Swezey CS, Seefert EC, Parker M, 2018, "A Brief Geological History of Cockspur Island at Fort Pulaski National Monument, Chatham County, Georgia," United States Geological Survey Fact Sheet 2018–3011.

6. Swezey et al., 2018. This formation was deposited between 3.5 and 12 million years ago during the Miocene and Pliocene epochs.

7. Lee likely concluded that the fall of the four sand forts at Hatteras Inlet and Port Royal indicated that temporary fortifications on barrier islands were overvalued and resistance was more likely to be successful farther inland.

8. This discussion of Fort Pulaski focuses on the effectiveness of rifled artillery against brick structures. Chapter 15 discusses the construction, history, and siege to capture the fort in more detail.

9. Lattimore RB, 2015, *Fort Pulaski National Monument—Georgia* (Washington, D.C.: U.S. Government Printing Office), 65p.

10. Lattimore R.B, 1954, *Fort Pulaski National Monument, Georgia*, Historical Handbook Series, no. 18 (Washington, D.C.: National Park Service).

11. Page D 1994, *Ships versus Shore: Civil War Engagements along the Southern Shores and Rivers* (Nashville: Rutledge Hill), p. 159.

12. Dougherty K, 2010, *Strangling the Confederacy: Coastal Operations in the American Civil War* (Havertown, Pa.: Casemate), p. 65.

13. Gillmore QA, 1865, *Official Report to the United States Engineer Department of the Siege and Reduction of Fort Pulaski, Georgia, February, March and April 1862* (New York: D. Van Nostrand), https://lccn.loc.gov/06038857.

14. See John Weaver's *A Legacy in Brick and Stone* for a detailed description of the construction, and reduction, of Fort Pulaski. Weaver JR, 2018, *A Legacy in Brick and Stone: American Coastal Defense Forts of the Third System, 1816–1867* (McLean, Va.: Redoubt), 329p.

15. Page, 1994, p. 159.

16. Chaitin PM, 1984, *The Coastal War: Chesapeake Bay to Rio Grande* (New York: Time-Life), 176p., quotation on p. 49. Porter was quite fond of the bullets-as-insects trope. He wrote of his experience under Sheridan at Five Forks: "Bullets were humming like a swarm of bees." Quoted in Kurtz HI, 2006, *Men of War: Essays on American Wars and Warriors* (Bloomington, Ind.: Xlibris), quotation on p. 161.

17. This reassignment certainly seemed to ignore his background in coastal engineering and his recent tactical success on barrier islands.

18. Duke University Press offers a series of state-specific coastal books with good explanations of the general geomorphology and geological history of the Atlantic and Gulf coast barrier islands. These books are edited by two well-respected coastal geologists: Orrin Pilkey and William Neal.

19. Federal artillery took full advantage of barrier island sediment when positioning their guns during the siege of Fort Macon, as demonstrated by this report from Brig. Gen. John G. Parke: "In selecting sites for our batteries advantage was taken of the sand hills previously spoken of. By cutting down the

natural slopes of these hills to a sufficient depth to lay the platforms for our guns and mortars and revetting the interior spaces with sand bags excellent epaulements were formed. Embrasures for the Parrott guns were cut directly through the sand hills, and revetted with sods taken from the salt-marsh close at hand." *The War of the Rebellion: A Compilation of the Official Records of the Union and Confederate Armies*, ser. 1, v. 9, quotation on p. 284.

20. The exact timing of the addition of sand to the scarp of Moultrie is something of a mystery—even the source of the sand deposition (anthropogenic vs. hurricane deposition) is in doubt. The sand may have played a role in Maj. Robert Anderson's decision to move his small garrison from Moultrie to Fort Sumter. The sand piles against the side of the fort might allow hostile townsfolk to flood over the walls of the fort and quickly overwhelm his men.

21. The fortification was only called the Neck Battery for a short period of time. During later Confederate occupation, it was renamed Battery Wagner. Even later, after Union capture, the northern face of the fort was strengthened and the works were referred to as Fort Strong. This last name is rarely used in the literature.

22. Hippensteel SP, 2010, "Barrier Island Geology and Union Strategy for the Assault and Siege of Charleston, South Carolina, 1862–1863," *Southeastern Geology*, v. 48, n. 1, pp. 23–35. Technically, the Neck Battery was not a fort; forts have protection from ramparts or parapets (or other earthworks) on all faces, while batteries offer only a limited degree of protection to the rear. Forts are built to withstand attack from all sides; batteries are built to withstand attack only from the front. After Federal occupation and the addition of a new parapet facing Charleston Harbor the fortification became a proper fort.

23. Hippensteel, 2010, p. 24.

24. Oertel GF, 1985, "The Barrier Island System," *Marine Geology*, v. 63, pp. 4101–18.

25. Belknap DF, Kraft JC, 1985, "Influence of Antecedent Geology on Stratigraphic Preservation Potential and Evolution of Delaware's Barrier Systems," *Marine Geology*, v. 63, pp. 235–62.

26. Oertel, 1985.

27. Harris MS, Gayes PT, Kindinger JK, Flocks JG, Krantz DE, Donovan P, 2005, "Quaternary Geomorphology and Modern Coastal Development in Response to an Inherent Geologic Framework: An Example from Charleston, South Carolina," *Journal of Coastal Research*, v. 21, pp. 49–64.

28. Cote R, 1985, *Jewel of the Cotton Fields: A History of Secessionville Manor* (Mt. Pleasant, S.C.: Richard N. Cote); Trinkley M, Hacker D, 1997, *Excavations at a Portion of the Secessionville Archaeological Site (38ch1456), James Island, Charleston County, South Carolina* (Columbia: Chicora Foundation), 201p.

29. The presence of this formidable—if somewhat smaller—fort, and to a lesser extent Fort Sumter, diminished the naval support available for a strike against Charleston from the north.

30. Hippensteel, 2010.

31. Contrary to what most people think, the first shot of the Civil War, fired at Fort Sumter, was launched from Fort Johnson, not Fort Moultrie.

32. Bostick DW, 2010, *Charleston under Siege: The Impregnable City* (Charleston, S.C.: History Press), 160p.

33. Wise SR, 1994, *Gate of Hell: Campaign for Charleston Harbor, 1863* (Columbia: University of South Carolina Press), 312p.

34. Ripley W, 1996, *The Civil War at Charleston* (Charleston, S.C.: News and Courier/Evening Post), 85p., p. 35.

35. Charleston ebb and flood tides typically run at around 3 knots.

36. https://markerhunter.wordpress.com/tag/maffitts-channel/ (accessed June 20, 2019).

37. U.S. Coastal Survey: Map of Comparison of Maffitt's Channel, 1855, 1:5000 scale (Hydrographic Party under the Command of J. N. Maffitt).

38. The *New Ironsides* had a flat keel, explains Bostick (2010), making the ironclad especially difficult to control in the strong currents. As the ship came to a halt on the fringe of the channel, she was sitting directly above an electronically controlled Confederate torpedo, which failed to explode. The *New Ironsides* nearly gave the harbor a new iron bottom.

39. Bostick, 2010, p. 65.

40. Bostick, 2010, p. 65.

41. Weaver, 2018, pt. 1.

42. James W. Hagy (1993) gives a detailed history of the military operations and living conditions on Folly Island during the second half of the Civil War. *To Take Charleston: The Civil War on Folly Island* (Missoula, Mont.: Pictorial Histories), 88p.

43. The fortification of Little Folly Island is further discussed in chapter 14.

44. Wise, 1994, p. 67.

45. Wise, 1994, p. 70.

46. Horres CR, Jr, 2019, *Morris Island and the Civil War: Strategy and Influence* (Charleston, S.C.: History Press), 144p.

47. Wise, 1994, p. 71.

48. Wise, 1994, p. 76.

49. Hippensteel, 2010.

50. Emilio LF, 1894, *History of the Fifty-Fourth Regiment of Massachusetts Volunteer Infantry, 1863–1865* (Boston: Boston Book Company), 452p.

51. Horres, 2019, p. 90.

52. Gillmore QA, 1865, *Engineer and Artillery Operations against the Defences of Charleston Harbor in 1863: Comprising the Descent upon Morris Island, the Demolition of Fort Sumter, the Reduction of Forts Wagner and Gregg; with Observations on Heavy Ordnance, Fortifications, etc.* (New York: D. Van Nostrand), 354p.

53. Bostick, 2010, pp. 119–21.

54. Confederate sharpshooters in and around Battery Wagner had received

several Whitworth rifles from Charleston. The Whitworth was the longest-ranging and most accurate sniper rifle from the Civil War, and it was especially effective when modified with an optical scope.

55. Horres, 2019, p. 92.

56. Horres (2019, p. 16) documents another "first" on Morris Island: "The range of the rifled artillery exceeded the ability of the eye to see the point of impact, and the first recorded use of compass-directed artillery fire was from Morris Island at the city of Charleston during the campaign."

57. According to Noah Trudeau (2008), pontoon trains became Sherman's "secret weapon" as he crossed Georgia toward Savannah. The rivers and streams that had slowed so many campaigns in the years before his march were now much less of an obstacle under the leadership of the dedicated engineer and pioneer. *Southern Storm: Sherman's March to the Sea* (New York: Harper Perennial), 688p.

58. Bostick, 2010, p. 139.

59. National Park Service, https://www.nps.gov/fosu/learn/historyculture/quincy-a-gillmore.htm (accessed June 1, 2019); Bostick, 2010, p. 92; Horres, 2019, pp. 93–95.

60. *War of the Rebellion: A Compilation of the Official Records of the Union and Confederate Armies*, ser. 1, v. 28, p. 27.

61. *War of the Rebellion: A Compilation of the Official Records of the Union and Confederate Armies*, ser. 1, v. 28, p. 27.

62. Recent studies by the U.S. Army Corps of Engineers indicates that the non-marsh sediment on Tybee Island is almost entirely composed of fine and medium-fine sand. USACE, April 2019, *Draft Environmental Assessment and Finding of No Significant Impact Tybee Island, Georgia Shoreline Protection Project 2019 Hurricane Harvey, Irma, Maria Emergency Supplemental Renourishment*, 81p.

63. Gillmore QA, 1865, *Official Report to the United States Engineer Department of the Siege and Reduction of Fort Pulaski, Georgia, February, March and April 1862* (New York: D. Van Nostrand). The angle of repose for dry, fine sand is just over 30°. This angle increases with increasing grain size and decreases with increasing roundness of the grains. If the sediment is damp, the angle of repose can reach 45°.

64. In Gillmore's supplemental report, titled *Engineering and Artillery Operations against the Defenses of Charleston Harbor in 1863*, he makes forty-nine references to "sand," a similar word count to terms like "Wagner" or "rifle."

65. *War of the Rebellion: A Compilation of the Official Records of the Union and Confederate Armies*, ser. 1, v. 28, p. 28.

66. Gillmore, 1865.

CHAPTER 16. The Strength of Sand

1. World War I saw the first widespread use of geologists as scientists in a military setting.

2. See Brown A, 1963, "A Geologist-General in the Civil War," *Geotimes*, v. 7, no. 7, pp. 8–11.

3. The soil is also, of course, critical for agriculture. Over the last decade many environmental historians have published texts investigating the link between soil type (and depletion and erosion) and the causes of the war. For example, Mark Fiege (2013) devoted a chapter to "The Nature of Gettysburg," where he discusses the "biophysical resources" available to each side of the conflict, and how soils and minerals of each region influenced both the causes, and outcome, of the war. *The Republic of Nature: An Environmental History of the United States* (Seattle: University of Washington Press).

4. In the field, sedimentary geologists differentiate mud (silt + clay) from pure clay by chewing the sediment: clay is smooth but the silt in mud makes it chew with a "gritty" texture.

5. The perfect material for inclusion in a sandbag.

6. Brig. Gen. Thomas Williams, while attempting to dig a canal through the Mississippi River's great floodplain, attempted to use the river's natural erosive power to his benefit. He found this could not always be accomplished because of the cohesion of the sediment, writing of his frustration: "The current of the river, however great, will not wash away the clay." *The War of the Rebellion: Formal Reports, Both Union and Confederate, of the First Seizures of United States Property in the Southern States, and of All Military Operations in the Field, with the Correspondence, Orders and Returns Relating Specially Thereto*, 1880–1898, v. 1-53 [serial no. 1-111], 111v, p. 27.

7. If the fortification is constructed of clay, particularly wet clay, penetration will be much easier.

8. Sand in the path of the shell also grinds together in front of the projectile, creating heat and friction between the impacted and compressed grains; this will also help diminish velocity.

9. Not only is repair relatively easy, there were numerous reports in the *Official Records* that made reference to the self-healing nature of sandy parapets. The sediment was tossed in the air after an explosion, only to fall harmlessly back into its original location.

10. Hippensteel SP, 2010, "Barrier Island Geology and Union Strategy for the Assault and Siege of Charleston, South Carolina, 1862–1863," *Southeastern Geology*, v. 48, n. 1, pp. 23–35.

11. Cohesion was also a great problem for the canal diggers around Vicksburg. The diaries and reports of the soldiers are full of descriptive terms relating to the difficulty of working with cohesive fine sediment: "stiff clay," "tenacious" clay soil. *The War of the Rebellion: A Compilation of the Official Records of the Union and Confederate Armies*, ser. 1, v. 24, pp. 119 and 65.

12. Bailey C, 2014, "The Saddest Affair: A Geological Perspective on the Battle of the Crater, U.S. Civil War," *W&M Blogs*, www.wmblogs.wm.edu.

13. Michael C. C. Adams (2016) devotes a chapter of *Living Hell: The Dark Side*

of the Civil War (Baltimore: John Hopkins University Press) to the effects of mud and dust on soldiers' morale.

14. Examples of "miraculous" escapes from exploding heavy ordnance by men surrounded by sand are numerous. For example, Maj. Lewis Arnold, First U.S. Artillery, remarked about one circumstance from the bombardment of Pensacola: "The fire from the enemy's batteries was heavy and well directed. There were many marvelous escapes from wounds. Among the most notable was that of Lieutenant Shipley, Third Infantry, and the detachment serving the 10-inch columbiad *en barbette* of his battery. A 10-inch shell struck the shell-proof and burst among his men and himself without wounding any one, although the sand and sand bags were knocked down over and around them." *The War of the Rebellion: A Compilation of the Official Records of the Union and Confederate Armies*, ser. 1, v. 6, p. 475.

15. The stratigraphy of the back-barrier marshes behind Folly and Morris Islands is discussed in chapter 18.

16. See R. S. Durham (2008) for a detailed history of the construction and combat history of this earthen fortification. *Guardian of Savannah: Fort McAllister, Georgia, in the Civil War and Beyond* (Columbia: University of South Carolina Press), 282p.

17. Department of the Air Force, Air Force Handbook 10–222, v. 14, 2008, *Civil Engineering Guide to Fighting Positions, Shelters, Obstacles, and Revetments*, 157p.

18. Department of the Air Force, 2008.

19. Material penetration has always confounded experts on terminal ballistic effects, even to the present day. Nonmilitary hollow-point bullets are designed to expand when hitting a soft target in a manner that will transfer as much energy as possible to the target, creating a larger wound cavity. Nevertheless, a thick winter coat has been known to cause premature expansion, limiting the effectiveness of pistol ammunition.

20. Department of the Air Force, 2008.

21. Demonstrations of Civil War artillery are common on Revolutionary War and Civil War battlefields today. Most include firing of "blank" charges—a powder charge without a round of actual ammunition. One of the more startling observations I made at Malvern Hill battlefield during one such reenactment (with live ordnance!) was just how easy it was to see and track a Civil War mortar shell in the air. The mortar probably wasn't fired at full velocity, but I could easily trace the path of the round as it was sent downrange, and anyone standing near where it landed would probably have been able to follow its path as well, giving them a few seconds to avoid it or take cover.

22. Note that the A-10 Warthog was developed *after* 1865—tank-killing aircraft were rarely seen on the battlefields of the Civil War.

23. Mahan DH, 1862, *A Treatise on Field Fortification, Containing Instructions on the Methods of Laying Out, Constructing, Defending, and Attacking Intrenchments; with the General Outlines Also of the Arrangement, the Attack and Defense of Permanent Fortifications* (Richmond, Va.: West & Johnson), 258p.

24. Mahan, 1862, p. 19.

25. Mahan, 1862, p. 8.

26. Gillmore Q, 1890, *The War of the Rebellion: A Compilation of the Official Records of the Union and Confederate Armies*, ser. 1, v. 28, pt. 1, pp. 27–28.

27. Gillmore, 1890, p. 27.

28. Gillmore, 1890, pp. 27–28.

29. Brooks TB, 1890, *The War of the Rebellion: A Compilation of the Official Records of the Union and Confederate Armies*, ser. 1, v. 28, pt. 1, note 15.

30. Smaller-caliber, more pointed Sharps bullets would probably penetrate deeper than .577 caliber Enfield bullets, and Spencer bullets had a lower muzzle velocity.

31. *The War of the Rebellion: A Compilation of the Official Records of the Union and Confederate Armies*, ser. 1, v. 6, p. 336. Brooks was clear about his opinion of the utility of sandbags: "It is hard to conceive how this siege could have been conducted without sand-bags. Forty-six thousand one hundred and seventy-five, according to the account kept at the engineering depot, have been expended on the portion of the work herein described."

32. Gillmore, Q, 1890, *The War of the Rebellion: A Compilation of the Official Records of the Union and Confederate Armies*, ser. 1, v. 28, pt. 1, p. 35.

CHAPTER 17. Gibraltar of the South

1. This measurement is a composite estimate based on multiple bathymetry surveys conducted from the 1840s and 1850s. Smithville was renamed Southport after the war in hopes of attracting more oceangoing commerce.

2. This coastal geomorphology is not unlike that of Cockspur Island on the Savannah River, the site of Fort Pulaski.

3. Fonvielle C, 2001, *The Wilmington Campaign: The Last Rays of Departing Hope* (Shippensburg, Pa.: Stackpole), 623p.

4. Pilkey OH, Neal WJ, Riggs SR, Webb CA, Bush, DM, 1998, *The North Carolina Shore and Its Barrier Islands: Restless Ribbons of Sand* (Durham, N.C.: Duke University Press), 344p.

5. Riggs SR, Ames DV, Culver SJ, Mallinson DJ, 2011, *The Battle for North Carolina's Coast: Evolutionary History, Present Crisis, and Vision for the Future* (Chapel Hill: University of North Carolina Press), 142p., discussion of shoals on pp. 36–38.

6. Trotter WR, 1989, *Ironclads and Columbiads* (Winston-Salem, N.C.: John F. Blair), 456p.

7. U.S. Coast Survey Map from 1851, Survey by Lieuts. Jenkins and Maffitt.

8. Moore MA, 1999, *Moore's Historical Guide to the Wilmington Campaign and the Battles for Fort Fisher* (Mason City, Iowa: Savas), p. 7.

9. Weaver JR, 2018, *A Legacy in Brick and Stone: American Coastal Defense Forts of the Third System, 1816–1867* (McLean, Va.: Redoubt), p. 200.

10. J. Taylor Wood's letter to Jefferson Davis clearly demonstrates his opinion of the comparative defensive strength of Forts Fisher and Caswell: "Fort Fisher,

at New Inlet, is a series of sand and palmetto works, which with proper weight of metal, could defy any water attack. Fort Caswell, much weaker, is in a transition state; the masonry as far as possible being covered with sand, and on two faces of the work an inclined shield covered with railroad iron and sand-bags is being erected." *The War of the Rebellion: A Compilation of the Official Records of the Union and Confederate Armies*, ser. 1, v. 51, quotation on p. 681.

11. The Southerners unsurprisingly changed the name of this peninsula to "Confederate Point."

12. Much earlier in the war Braxton Bragg had reported another way that sand could hinder military efforts, acting as the opposite of a force multiplier: "Our loss is more severe than at first reported. The men became much exhausted from the long and fatiguing march through the deep sand of the island, and no doubt a considerable portion of the loss was from this cause." Report from Pensacola Florida, October 10, 1861, in *The War of the Rebellion: A Compilation of the Official Records of the Union and Confederate Armies*, ser. 1, v. 6, quotation on p. 470.

13. Moore, 1999, p. 13.

14. Trotter, 1989, p. 327.

15. The Mound Battery towered 60 feet above the high-tide line.

16. Moore, 1999.

17. Trotter, 1989, p. 328.

18. These land mines contained approximately 100 pounds of black powder.

19. Similar post-construction modifications were made to Battery Wagner on Morris Island. After capture, the Federal engineers quickly added a new sally port directly through the original land face. The fort was essentially rotated 180°, with the new defensive land face directed toward the north and Charleston Harbor.

20. A demilune is a half-moon–shaped earthwork that exists outside of the fortification's main walls. It allows defenders to fire along the face of a fortification into the flank of an attacker. If the earthwork is triangular in shape, it is called a ravelin.

21. Welles G, 1911, *Diary of Gideon Welles, Secretary of the Navy under Lincoln and Johnson, with an Introd. by John T. Morse* (Boston: H. Mifflin), 653p., quotation on p. 127.

22. Barratt P, 2004, *Circle of Fire: The Story of the USS Susquehanna in the War of the Rebellion* (Bromley, U.K.: Columbiad Press), 245p.

23. Fonvielle, 2001, p. 137.

24. A more detailed discussion of these bombardments, and the effects of the shelling on the sandy parapets, can be found in Gragg R, 2006, *Confederate Goliath: The Battle of Fort Fisher* (Baton Rouge: Louisiana State University Press), 400p.

25. Gragg, 2006, quotation on p. 122.

26. *The War of the Rebellion: A Compilation of the Official Records of the Union and Confederate Armies*, ser. 1, v. 46, quotation on p. 408.

27. Moore, 1999, p. 43.

28. Trotter, 1989, pp. 393–96. Around a hundred of the men were also armed with Spencer repeating rifles. These men were primarily used as skirmishers and sharpshooters. The arming of the sailors and marines with only revolvers and sabers is one of the more curious decisions of the campaign. This tactic must have been borne from a belief that the strike force would move more quickly and aggressively toward the parapets of the fort if only lightly armed. With rifles, the men might have been tempted to fire at range and reload, robbing them of precious momentum and giving the Confederate garrison time to prepare for the beachfront assault.

29. Moore, 1999, pp. 70–71.

30. Sesquicentennial Spotlight: The Fall of Fort Fisher and the Confederacy's Collapse, http://www.civildiscourse-historyblog.com/blog/tag/117th+New+York+Infantry (accessed January 12, 2020).

31. Hippensteel S, Eastin M, Garcia W, 2013, "The Geologic Legacy of Hurricane Irene: Implications for the Fidelity of the Paleo-storm Record," *GSA Today*, v. 23, n. 10, pp. 4–10.

32. Moore, 1999, p. 149.

33. Lamb W, 1893, *Fort Fisher, The Battles Fought There in 1864 and 1865*, Southern Historical Society Papers, v. 21 (Richmond: Virginia Historical Society), pp. 257–90.

CHAPTER 18. Sedimentary Geology as a Tool for History

1. Kraft J, Rapp G, Szemler G, Tziavos C, Kase E, 1987, "The Pass of Thermopylae, Greece," *Journal of Field Archaeology*, v. 14, pp. 181–98.

2. Vouvalidis K, Syrides G, Pavlopoulos K, Pechlivanidou S, Tsourlos P, Papakonstantinou MF, 2010, "Palaeogeographical Reconstruction of the Battle Terrain in Ancient Thermopylae, Greece," *Geoeodinamica Acta*, v. 23, pp. 241–53.

3. Thoms AV, 2004, "Sand Blows Desperately: Land-use History and Site Integrity at Camp Ford, a Confederate POW Camp in East Texas," *Historical Archaeology*, v. 38, pp. 73–95; Hippensteel SP, 2008, "Reconstruction of a Civil War Landscape: Little Folly Island, South Carolina," *Geoarchaeology*, v. 23, n. 6, pp. 824–41.

4. On some topographic maps Rat Island is depicted as being the extension of the northern portion of Little Folly Island where it extends into the back-barrier marsh. Rat Island is a small rise of ground behind Folly and along the edge of Lighthouse Inlet.

5. Zierden MA, Smith SD, Anthony RW, 1995, *"Our Duty Was Quite Arduous": History and Archaeology of the Civil War on Little Folly Island, South Carolina*, Charleston Museum Leaflet Series, no. 32 (Charleston: Charleston Museum).

6. Hippensteel, 2008.

7. Foraminifera are single-celled creatures that have either a calcareous or agglutinated test (internal shell). Agglutinated varieties use organic cement to glue silt and sand particles together to make their test.

8. Hippensteel SP, 2011, "Spatio-Lateral Continuity of Storm Overwash Deposits in Back Barrier Marshes," *Geological Society of America Bulletin*, v. 123, pp. 2277–94.

9. Lennon G, Neal WJ, Bush DM, Pilkey OH, Stutz M, Bullock J, 1996, *Living with the South Carolina Coast* (Durham, N.C.: Duke University Press), 264p.

10. Several other examples of Civil War archaeological studies can be found in Scott D, Babits L, Haecker C, eds., 2009, *Fields of Conflict: Battlefield Archaeology from the Roman Empire to the Korean War* (Dulles, Va.: Potomac).

11. Sedimentary structures are large-scale features like mud cracks or graded bedding that are incorporated into a sediment before it lithifies. These structures are valuable for identifying original depositional environments or facies.

12. Hippensteel, 2008.

13. Solly M, 2019, "Hurricane Dorian Unearths Civil War Cannonballs at South Carolina Beach," *Smithsonian*, https://www.smithsonianmag.com/smart-news/hurricane-dorian-unearths-civil-war-cannonballs-south-carolina-beach-180973095/ (accessed December 2, 2019).

14. These sediments often contain Oligo-Miocene taxa of foraminifers, indicating they were reworked (eroded) from ancient outcrops of sediment on the continental shelf.

15. Jacobsen M, Downs JCU, Harris MS, Scafuri M, Owsley D, Abrams LC, Rieders M, Gregory RD, Mardikian P, Hippensteel S, 2013, "Forensic Investigations of the American Civil War Submarine *H. L. Hunley*," American Academy of Forensic Sciences, 65th Annual Scientific Meeting, February 18–23, 2013, Washington, D.C.

16. Harris MS, Neyland RS, 2016, "Environmental Context and Site Formation," pp. 51–65 in Neyland RS, Brown H, eds., *H. L. Hunley Recovery Operations* (Washington, D.C.: Naval History & Heritage Command, Department of the Navy), 321p.

17. Taphonomy is the study of what happens to a creature between the time of death and the time of discovery by a paleontologist or archaeologist (or, really, anyone else). It includes analysis of burial conditions, mode of preservation, and physical and chemical alteration of the fossil.

18. Jacobsen et al., 2013.

19. Hippensteel SP, 2007, "The Role of Foraminiferal Analysis in the Interpretation of *H. L. Hunley*'s Site Formation Processes," in SHA, 2007, *Old World/New World: Culture in Transformation, Society for Historical Archaeology*, abstract, p. 266. Williamsburg, Va.

20. Mallon J, Glock N, Schönfeld J, 2011, "The Response of Benthic Foraminifera to Low-Oxygen Conditions of the Peruvian Oxygen Minimum Zone," in *Anoxia: Paleontological Strategies and Evidence for Eucaryotic Survival* (New York: Springer, Dordrecht), pp. 305–21.

21. Auer A, Mottonen M, 1988, "Diatoms and Drowning," *Zeitschrift fur Rechtsmedizin*, v. 101, pp. 87–98.

22. A few years ago I was approached by the Charlotte-Mecklenburg County sheriff's office to identify the diatoms recovered from the hair of a young woman whose corpse was dumped under a railroad bridge. The detectives were interested in knowing if the diatoms might indicate from where the body had been transported before being discarded in south Charlotte.

23. When soft tissue in corpses is exposed to water for long periods of time, it may transform into a waxy, gray/yellowish substance known as adipocere.

24. Auer and Mottonen, 1988; Hürlimann J, Feer P, Elber F, Niederberger K, Dirnhofer R, Wyler D, 2000, "Diatom Detection in the Diagnosis of Death by Drowning," *International Journal of Legal Medicine*, v. 114, pp. 6–14; Sidari L, Di Nunno N, Costantinides F, Melato M, 1999, "Diatom Test with Soluene-350 to Diagnose Drowning in Sea Water," *Forensic Science International*, v. 103, pp. 61–65; Takeichi T, Kitamura O, 2009, "Detection of Diatom in Formalin-Fixed Tissue by Proteinase K Digestion," *Forensic Science International*, v. 190, pp. 19–23.

CHAPTER 19. The Fate of the Fortifications

1. Edward Tabor Linenthal (1993) provides an interesting history of the controversy surrounding the Gettysburg battlefield and the struggle between those who wish to return the fields to the appearance from 1863 and those who would rather allow the landscape to evolve and change through time. *Sacred Ground: Americans and Their Battlefields* (Champaign: University of Illinois Press), 352p.

2. Gallagher GW, 2020, *The Enduring Civil War: Reflections on the Great American Crisis* (Baton Rouge: Louisiana State University Press), 296p.

3. And certainly foot traffic.

4. The Railroad Redoubt and Texas Memorial are seriously threatened by an eroding bluff that separates these artifacts from the Kansas City Southern rail tracks. In July 2017 the National Park Service proposed building a soldier pile wall to limit further erosion and stabilize the loess slope, protecting the earthworks and monument (U.S. Department of Interior, Railroad Redoubt Stabilization Plan).

5. National Park Service, 1989, *Earthworks Landscape Management Manual* (Washington, D.C.: Park Historic Architecture Division, Cultural Resources U.S. Department of the Interior).

6. Aust WM, Azola A, Johnson JE, 2003, "Management Effects on Erosion of Civil War Military Earthworks," *Journal of Soil and Water Conservation*, v. 58, pp. 13–20.

7. National Park Service, https://www.nps.gov/tps/how-to-preserve/currents/earthworks/introduction.htm (accessed January 8, 2020).

8. Ibid.

9. Archaeological Society of Virginia, https://www.archeologyva.org/PDF/HICKS-TIMECRIME.pdf (accessed November 12, 2019).

10. To be classified as a "barrier island," a unit of land must be surrounded on all sides by water: two inlets separate the island from adjacent land (or other islands), the ocean lies on one front, and either a marsh or lagoon lies between the island and the mainland. Islands in environments where sediment is abundant will usually have a marsh or mudflats behind them, depending on the richness of vegetation, while islands in sediment-poor areas will have a lagoon.

11. Estimates vary greatly on the overall cost, depending on whether beach replenishment projects are considered in the overall total. This estimate is from the U.S. Global Change Research Program, which is managed by the National Oceanic and Atmospheric Administration. Beach replenishment, or beach nourishment, projects are becoming an increasingly popular option when compared with hard armoring or stabilization, to the point that the sediment being used in such projects is becoming scarce (and more expensive).

12. Coastal Maine is a good example of a shoreline underlain by igneous and metamorphic rocks.

13. See Pilkey OH, Pilkey-Jarvis L, 2007, *Useless Arithmetic: Why Environmental Scientists Can't Predict the Future* (New York: Columbia University Press, 230p.), for a discussion of the factors contributing to this complexity.

14. The rate of local sea-level rise is extremely high in regions that are subsiding because of groundwater or petroleum extraction. Louisiana and Mississippi have fortifications like Forts Jackson, St. Philip, and Massachusetts that are especially threatened by sea-level rise.

15. The shifting of inlets and the eventual erosion of coastal fortifications was identified as a potential problem even during the war. George B. McClellan included this report after the capture of Fort Hatteras: "There is a possibility that the cut between the two forts may soon be closed by the operation of natural causes, and if it remains open, the point on which Fort Hatteras stands may remain an island instead of being entirely washed away, as General Williams seems to anticipate. If it remains an island, I would suggest that measures be taken to prevent the washing away of its site in ordinary weather by means akin to those used in protecting the dikes of Holland." *The War of the Rebellion: A Compilation of the Official Records of the Union and Confederate Armies*, ser. 1, v. 4, quotation on p. 628.

16. This rate is most realistic for especially low-relief, gently sloping coastal regions like North and South Carolina.

17. Many "permanent" brick forts would be lost as well. Forts Morgan, Gaines, Pickens, Moultrie, Caswell, Macon, and Cinch are all within 1,000 feet of the actively eroding ocean.

18. Or, later, inlets that had been somewhat stabilized by jetties, terminal groins, or other armoring devices.

19. Although both Bogue Banks and more recently Cockspur Island have received some degree of anti-erosion armoring in the twentieth century.

20. Page D, 1994, *Ships versus Shore: Civil War Engagements along the Southern Shores and Rivers* (Nashville: Rutledge Hill), 410p.

21. Weaver JR, 2018, *A Legacy in Brick and Stone: American Coastal Defense Forts of the Third System, 1816–1867* (McLean, Va.: Redoubt), p. 205. This battery was part of the Endicott Period system of dispersed gun emplacements, where disappearing heavy artillery was positioned behind concrete walls. Fort Moultrie also has several similar open-topped batteries, although none are located directly in the fort.

22. Pilkey OH, Neal WJ, Riggs SR, Webb CA, Bush DM, 1998, *The North Carolina Shore and Its Barrier Islands: Restless Ribbons of Sand* (Durham, N.C.: Duke University Press), 344p.

23. Pilkey et al., 1998, p. 191.

24. Gillmore QA, 1884, "Report of the Chief of Engineers," in *Annual Report of the Secretary of War*, v. 2 (Washington, D.C.: Government Printing Office), 1,080p.

25. The direction of beach drift can be easily determined for beaches that have been "stabilized" or armored with shore-perpendicular groins. Drift is occurring from the side of the groin that has more trapped sand toward the side with less.

26. The USS *Housatonic*, the first ship ever sunk by a submarine, was sitting on a shallow portion of this delta. When it went to the bottom in the shallow water, most of the crew simply climbed into the rigging to await rescue.

27. Lennon G, Neal WJ, Bush DM, Pilkey OH, Stutz M, Bullock J, 1996, *Living with the South Carolina Coast* (Durham, N.C.: Duke University Press), 264p.

28. Minimal wind-driven water movement will occur at water depths greater than half the wavelength of the waves. Larger storm waves could probably still move the sand, however.

29. On a more positive note, some of these redirected sediments helped bury and preserve the Confederate submarine *H. L. Hunley*.

30. Coleman JC, 1988, *Fort McRee: "A Castle Built on Sand"* (Pensacola, Fla.: Pensacola Historical Society), 124p.

31. Muir T Jr, Ogden DP, 1989, *The Fort Pickens Story* (Pensacola, Fla.: Pensacola Historical Society), 24p.

32. Field R, 2007, *American Civil War Fortifications (3): The Mississippi and River Forts* (Oxford: Osprey), p. 57.

33. The earthworks at Grand Gulf have suffered from severe cratering from unusual sources. The Confederate garrison purposely detonated the main magazine of Fort Wade before abandoning the site. Much more recently, tornadoes knocked over hundred-year-old trees, ripping up large craters from their uplifted roots.

CHAPTER 20. Lessons Learned for Future Conflict

1. Brady LM, 2012, *War upon the Land: Military Strategy and the Transformation of Southern Landscapes during the American Civil War* (Athens: University of Georgia Press), p. 67.

2. The South apparently had two Gibraltars, neither of which was composed of rock. The Vicksburg Gibraltar was strong because of loess, and the Cape Fear Gibraltar because of sand.

3. Zen E, Walker A, 2000, *Rocks and War: Geology and the Civil War Campaign of Second Manassas* (Shippensburg, Pa.: White Mane), 109p.

4. Bailey RH, 1984, *The Bloodiest Day: The Battle of Antietam* (New York: Time-Life), p. 154.

5. National Park Service, 2015, Kennesaw Mountain National Battlefield Park Georgia, Battlefield Pamphlet, U.S. Department of the Interior.

6. Even so, clay from the weathered feldspar-rich igneous rocks probably influenced the fighting to a greater degree: whenever possible, trenches, parapets, and even tunnels were constructed around the mountain.

7. Lewis ER, 1970, *Seacoast Fortifications of the United States: An Introductory History* (Annapolis, Md.: Naval Institute Press), 146p.

8. Weaver JR, 2018, *A Legacy in Brick and Stone: American Coastal Defense Forts of the Third System, 1816–1867* (McLean, Va.: Redoubt), p. 5.

9. Lewis (1970) explains that the key benefit of land-based artillery over guns on ships is accuracy (not necessarily firepower). With precise fire, one portion of a fort's ramparts could be repeatedly targeted until it was breached. More inaccurate naval gunfire, shooting from an unsteady platform, would spread shell hits, thus requiring more strikes before a rampart might eventually be pierced.

10. Gillmore QA, 1890, *The War of the Rebellion: A Compilation of the Official Records of the Union and Confederate Armies*, ser. 1, v. 28, pt. 1.

11. Gillmore QA, 1865, *Engineering and Artillery Operations against the Defenses of Charleston Harbor: Comprising the Descent upon Morris Island, the Demolition of Fort Sumter, the Reduction of Forts Wagner and Gregg; with Observations on Heavy Ordnance, Fortifications, etc.* (New York: D. Van Nostrand), 354p.

12. Weaver, 2018, p. 14, quotation on p. ix.

13. Iron "Totten shutters" had been added to the embrasures of some Third System forts, offering protection for the gun crews from return fire.

14. Wilson EE, 1920, *Notes on Seacoast Fortification Construction*, U.S. Army Engineering School, Occasional Papers Number 61, quotations on p. 178.

15. Lewis, 1970, p. 69.

16. Kaufmann JE, Kaufmann HW, 2007, *Fortress America: The Forts That Defended America, 1600 to the Present* (Cambridge, Mass.: Da Capo Press), p. 328. Rosendale cement was a weaker form of concrete compared to Portland cement, and the U.S. Army switched construction preference to the latter in 1899. With Rosedale cement the sand/concrete ratio was 2:1; with Portland cement it became roughly 3:1.

17. A strike from an artillery shell during the First World War would propel sediment a greater distance away from the crater, and the intensity of thousands of impacts could flatten earthworks in a short period of time. In the Civil War sandy parapets were rearranged by shellfire; in World War I they were obliterated, spreading the sediment across no-man's-land.

18. Longer, streamlined, aerodynamic, high-velocity bullets also lose less energy downrange and are less affected by crosswinds; as a result, they are significantly more accurate at longer ranges when compared to a blunt, broad .58-caliber minié ball.

19. For the most part; most history books point to the machine gun and artillery of the First World War as providing the coup de grâce for this offensive tactic.

20. Stephen Bull (2010) includes a chapter detailing the timing of the first extensive use of trenches in 1914. *Trench: A History of Trench Warfare on the Western Front* (Oxford: Osprey), 288p.

21. Murray N, 2013, *The Rocky Road to the Great War: The Evolution of Trench Warfare to 1914* (Lincoln: University of Nebraska Press), 320p.

22. Rose T, 2013, "Wartime Geotechnical Maps," *Geoscientist*, v. 23, pp. 10–15; Kaye CA, 1957, "Military Geology in the United States Sector of the European Theater of Operations during World War II," *Bulletin of the Geological Society of America*, v. 68, pp. 47–54.

23. Kaye, 1957.

24. Smith JSC, 1964, "Military Applications of Geology," *Transactions of the Kansas Academy of Sciences*, v. 67, pp. 311–36.

25. Kaye, 1957.

26. A DUKW, or "duck," is a six-wheel-drive 2.5 ton amphibious truck. As military vehicles they originally played an important role in World War II and the Korean War, but today they are best known for giving tours in cities such as Boston.

27. Lawson WD, 2008, "Soil Sampling at Sword Beach—Luc-Sur-Mer, France, 1943: How Geotechnical Engineering Influenced the D-Day Invasion and Directed the Course of Modern History," *International Conference on Case Histories in Geotechnical Engineering*, p. 13.

28. Historian David Howarth (1959) made this observation fifteen years after the D-Day landings. *D Day: The Sixth of June, 1944* (New York: McGraw-Hill), 251p.

29. The largest tidal range in Asia.

30. Rottman GL, Foster REM, 2007, *Vietnam Fortifications, Firebases and Tunnels* (Oxford: Osprey), 190p.

31. Knowles RB, Wedge WK, 1998, "Military Geology and the Gulf War," *Geological Society of America Reviews in Engineering Geology*, v. 13, pp. 117–24.

32. On asymmetrical sand dunes the windward slope of a dune, where wind has compacted the sand, might allow movement by a foot soldier or light vehicle. The lee slope, on the other hand, would avalanche because the sediment was near the angle of repose. See Knowles and Wedge (1998) for a detailed discussion of trafficability across desert landscapes.

33. Airborne sand, propelled at high velocity, can injure. From the journal of Maj. Gen. Jacob Cox, serving along the South Carolina coast: "The sand driving with the wind cuts like a knife and adds much to the unpleasantness of the

night." *The War of the Rebellion: A Compilation of the Official Records of the Union and Confederate Armies*, ser. 1, v. 47, pt. 1: Reports, quotation on p. 928.

34. Knowles and Wedge, 1998.

35. Brooks TB, 1863, *Official Records of the War of the Rebellion*, v. 28, no. 1, p. 266.

36. For good overviews of the defensive use of sandbags on Morris Island, S.C., see Hess EJ, 2005, *Field Armies and Fortification of the Civil War* (Chapel Hill: University of North Carolina Press), 448p; Wise SR, 1994, *Gate of Hell: Campaign for Charleston Harbor, 1863* (Columbia: University of South Carolina Press), 312p.

37. U.S. Army Corps of Engineers, Northwestern Division, 2004, *Sandbagging Techniques*, 8p.

38. And a clay-filled bag may turn into a large brick if the sediment becomes desiccated.

39. Brian Allen Drake devotes his chapter in Hersey MD, Steinberg T, eds., 2019, *Fields of Fire: The Future of Environmental History* (Tuscaloosa: University of Alabama Press), to making a similar argument for the study of environmental history during the Civil War.

40. Howarth, 1959.

41. Hupy J, Koehler T, 2011, "Modern Warfare as a Significant Form of Zoogeomorphic Disturbance upon the Landscape," *Geomorphology*, v. 157, pp. 169–82.

42. Bressan W, 2014, "The Geology of D-Day (June 6, 1944)," *Scientific American*, https://blogs.scientificamerican.com/history-of-geology/the-geology-of-d-day-june-6-1944/ (accessed February 12, 2020). Explosions leave behind more than craters: geologists Earle McBride (University of Texas at Austin) and M. Dane Picard (University of Utah) found that almost 4 percent of the sediment on Omaha Beach contained shrapnel forty-four years after the D-Day landings. This percentage was larger than the heavy mineral content in the fine-grained sand, exceeded only by quartz (78 percent) and feldspar (9 percent). McBride EF, Picard MD, 2011, "Shrapnel in Omaha Beach Sand," *Sedimentary Record*, v. 9, pp. 4–8.

43. Actually, the number exploded the year before (one-year residence time of cesium in the upper atmosphere).

44. This discrete spike in cesium has also been used to quantify the amount of bioturbation that has taken place in a layer of sediment. Sharp spikes indicate little mixing of the strata, while flattened peaks indicate that the radiotracer has been spread vertically into surrounding sediment layers by burrowing creatures.

45. Over fifty-eight years (1963–2021), 24 inches of rise = ~0.4 inches/year.

46. A second radiotracer, Pb-210, indicates that the infilling was largely complete many decades before this.

INDEX

The Families' Civil War: Northern African American Soldiers and the Fight for Racial Justice
BY HOLLY A. PINHEIRO JR.

Sand, Science, and the Civil War: Sedimentary Geology and Combat
BY SCOTT HIPPENSTEEL

CPSIA information can be obtained
at www.ICGtesting.com
Printed in the USA
LVHW090148150323
741658LV00014B/1053